William Forsell Kirby

Evolution and Natural Theology

William Forsell Kirby

Evolution and Natural Theology

ISBN/EAN: 9783337025953

Printed in Europe, USA, Canada, Australia, Japan

Cover: Foto ©berggeist007 / pixelio.de

More available books at **www.hansebooks.com**

EVOLUTION

AND

NATURAL THEOLOGY.

BY

W. F. KIRBY,

OF THE BRITISH MUSEUM.

LONDON:
W. SWAN SONNENSCHEIN & CO.,
PATERNOSTER ROW.
1883.

KELLY AND CO., PRINTERS, GATE STREET, LINCOLN'S INN FIELDS;
AND KINGSTON-ON-THAMES.

PREFACE.

A GREAT part of the present work was written some time ago, and an abstract of the philosophical portion was published as a series of papers in a London periodical. Want of leisure has hitherto prevented the author from preparing the completed manuscript for press. It now appears to him that his conclusions are still sound, and not altogether unworthy to be laid before the public, as the main argument remains essentially unaffected by publications more recent than those upon which the work was originally based.

The quotations from Darwin's "Origin of Species," are taken from the third edition, when not otherwise specified.

W. F. KIRBY.

London, 1883.

INTRODUCTION.

"I cannot but feel surprised that a theory which thus teaches us humility for the past, faith in the present, and hope for the future, should have been regarded as opposed to the principles of Christianity, or the interests of true religion."—LUBBOCK's "*Prehistoric Times*," 2nd edition, p. 581.

THE Theory of the Evolution of Living Beings, more familiarly known as the Theory of the Origin of Species, has continued to attract an increasing amount of public attention, ever since 1859, when Darwin published his great work on the subject. Although the principle of Evolution was not new, yet the crude and unscientific speculations of the earlier Evolutionists had failed to produce any deep or permanent impression on either the scientific or the popular mind; and it was left for Darwin and Wallace to promulgate a theory which could be seen to be both scientifically probable, and easily intelligible, and capable of accounting for a great number of familiar facts which had previously been regarded as lying almost beyond the domain of science, and therefore as incapable of explanation. It is universally acknowledged that no one can read one of Darwin's elaborate works on this subject, without admiring the great amount of learning and industry displayed in the marvellous array of facts collected from every conceivable source, which are brought forward to illustrate even the most trifling point under discussion.

The literature of the subject has now become very extensive, and Darwin's views are accepted, with more or less reservation, by nearly all scientific men, as a key to the mysteries of Nature. Among the supporters of Evolution may be found men of every shade of opinion, from Herbert Spencer, who asserts that Atheism, Pantheism and Theism are all equally untenable, to St. George Mivart, who attempts to show that the teachings of the Fathers of the Church are in accordance with Evolution.

But although the main principle of Evolution is now conceded by most naturalists, there is a great difference of opinion on matters of detail, and the subject is evidently still in its infancy. It must not be supposed that Darwin's views, comprehensive and valuable as they are, are by any means final; and every general work on Evolution attempts to develop the subject more extensively, or to throw light on certain questions which still remain obscure. It is often as important to point out the fallacies in the conclusions of others, as to state a new truth; for in science, as in arithmetic, a mistake in one figure will often affect the whole sum. There is little doubt that Evolution is still far from presenting the aspect which it will ultimately assume in the science of the future.

The religious aspects of Evolution, though frequently discussed, are still far from having received a satisfactory solution. One reason may be that pure science resembles pure mathematics in only being able to deal with subjects which fall directly within its grasp. Pure science will not conduct us far if the theological aspect of a theory has to be considered; for religious questions depend upon the constitution of the human mind, and we cannot call in science to enable us to analyse our own minds by their own powers from any really independent or scientifically unexceptionable standpoint. Nevertheless, a man is not justified in rejecting Religion, because (even after eliminating absurd or contradictory dogmas, with nothing to support them but authority) it presents him with insoluble enigmas if argued out by strict logic. A physiologist might as well refuse food as long as any problem connected with the process of digestion remained unsolved; or a philosopher might as well refuse to eat, because he admits himself to be incapable of proving either his own existence or that of his dinner. It is perhaps unnecessary to add that the above remarks are intended to apply to Religion in the abstract, quite independently of any system of dogmatic theology.

As, however, the bias of an author must affect his whole work, it may be stated that the existence and perfection of the Deity are here assumed from the outset, without any discussion of evidences, which would lead us too far from our main object. The chief points which we have endeavoured to establish are: (1) the worthlessness of the opinions of antiquity on matters of natural science; (2) the great superiority from a scientific point of view, as well as in physical and moral evidence, of the Theory of Evolution over that of Special Creation; and (3) that Evolution is perfectly consistent with an enlightened Theism. Some of the supporters of Evolution claim for it that it is wholly opposed to the belief in a God, and many of its opponents brand it as thoroughly Atheistic in its tendencies. We shall attempt to show that this is not the case: but that as the Theory of Special

Creation has become scientifically untenable, it is this theory, and not that of Evolution, which is positively Atheistic in its tendencies.

The principles of Evolution may be traced throughout every branch of Science without exception; but we are here principally concerned with that branch of the subject known as the Origin of Species, and our illustrations are taken chiefly from Zoology. The conclusions of previous authors (including those of Darwin himself) are freely criticised; and it is possible that new light may be found to be thrown on certain phases of the subject.

There is no real conflict between Religion and Science; it is only the worn-out theological beliefs of past ages which conflict with the latter, when the increased light of the present age demonstrates them to be false, or even immoral. Religion and Science both lead us on to the discovery of higher and higher truths; and the progress of truth is the advancement of the human race on the path which God has indicated by science and history as that by which it is His intention ultimately to lead all living beings, whether regarded as species or individuals, to Himself. The interests of truth are in all cases best promoted by the freest discussion; and it is hoped that the present attempt to throw light on some of the important religious and scientific questions of the day, from a somewhat unusual standpoint, may not be considered altogether uninteresting.

TABLE OF CONTENTS.

PAGE

PREFACE v

INTRODUCTION vii

CHAPTER I.

ANCIENT AND MODERN VIEWS OF NATURE.

The Universe as it appeared to the Ancients—Finite Nature of the Ancient Gods—Zeus—Othin—Elohim—Indra—The Universe as it appears to the Moderns—Progress of Science—Progress of Religion . . . 1

NOTES.—The Sacrifice of Serpents—Buddhism—God as the Spiritual Sun 13

CHAPTER II.

THEORY OF DIRECT CREATION.

Difference between Ancient and Modern Thought—Sporadic Nature of the Highest Genius—Ancient Opinions on Matters of Physical Science Worthless—Theory of Special Creation necessarily held by the Ancients—Miracles—Universality of Natural Law—The Witch-Mania and its Effects 15

CHAPTER III.

EVOLUTION IN ASTRONOMY AND PHILOLOGY.

Infinite Change throughout the Universe—Evolution an Established Principle in all Sciences—Astronomy—Philology—Analogy between Philological and Biological Evolution 30

Contents.

CHAPTER IV.

EARLIER THEORIES CONNECTED WITH EVOLUTION. PAGE

The System of Nature—Speculations of the Earlier Evolutionists—Theory of Degradation—Inspiration necessarily Opposed to Infallibility—Importance of Correct Views of Modern Science and Ancient Literature . . . 41

NOTE.—The Book of Genesis 49

CHAPTER V.

DARWIN AND HIS CRITICS.

Sketch of the Theory of Natural Selection—Objections—Theological Objections—Origin of Man—Beauty and Design in Nature—Mivart's Strictures on Darwinism—Sense-Organs—Origin of Species not always Gradual—Limits of Variability—Absence of Transitional Forms—Species and Races—Bree's Objections to Darwinism . 50

CHAPTER VI.

USE AND DISUSE, REVERSION AND HYBRIDISM.

Vital Energy—Its Limits—Variations in Offspring of the same Parents—Reversion—Hybrids and Mongrels—Effects of Domestication on Animals 69

CHAPTER VII.

HOMOLOGY.

Probable Unity of Matter—Unity of Structure throughout Organic Nature — Somites of Annulosa — Neck of Vertebrata—Rudimentary Organs—Organs of the Senses—Homology of the Sexes—Homology only to be explained by Evolution 78

CHAPTER VIII.

EMBRYOLOGY.

Importance of Embryological Characters in Classification—Close Resemblance of Early Stages of Embryos—Rudimentary Organs in the Embryo—Ancestral Characters Gradually Transferred to Embryo—Importance of this Principle 95

CHAPTER IX.

GEOGRAPHICAL DISTRIBUTION.

Traditions of the Deluge—Islands—Sea Barriers—Local Variations in Size of Animals—America and the Old World—Australia—Introduced Species—American Waterweed—Madagascar—The Canaries and the East Indies 103

CHAPTER X.

VARIATION UNDER DOMESTICATION.

Effects of Domestication—Sheep—Pigeons—Explanations on the Theory of Special Creation—Domestication in Ancient Times—Importance of Variation to Man . 119

CHAPTER XI.

ORIGIN OF LIFE ON THE EARTH.

Origin of Life a Mystery—Biassed Theories—Prochronism—Voltaire's Theory of Fossils—Sir W. Thomson's Meteoric Theory—Origin of Mundane Life probably Terrestrial—Lowest Living Organisms—Ultimate Origin of Life 126

CHAPTER XII.

COURSE OF DEVELOPMENT ON THE EARTH.

Succession of Forms—Theory of Direct Creation Subversive of the Argument from Design—Immutability and Universality of the Laws of Nature—Wallace's View of the Origin of Man—Absolute and Relative Perfection—Progress of Vertebrate Animals—Man Geologically Analogous to a Class—Extinction of Species—Small Species usually Survive Larger—Apparent Exceptions—Social Hymenoptera—Intelligence of Vertebrate Animals—The Dog—Indestructibility of Life—Pre-existence and Successive Existences 137

CHAPTER XIII.

THE DESTRUCTIVE AGENCIES OF NATURE. PAGE

Necessity of Destruction—Man's Influence—Primeval State of the Earth—Glacial Periods—Changes of Climate—Influence of Pestilence—Eras of Rest—State of Iceland—No Real Destruction 158

CHAPTER XIV.

PROGRESS OF MAN.

Alternation of Races—Children and Savages—Men only Equal Theoretically—Beliefs dependent on Race and Geographical Distribution—All Religions of Divine Origin—Missionaries—Hereditary Influences—Proportion of the Sexes—Alleged Degeneracy of Man—The Ancient Athenians—Celibacy in the Church—Cruelty and Immorality of the Ancients—Re-action in Modern Times 167

Notes.—Religion of the Semites and Aryans—Mohammadanism 199

CHAPTER XV.

HARMONY OF NATURE.

Balance of Nature—Strength of Small Animals—Limits of Variation—Of Multiplication—Heavy Fruits—Relative Perfection of the Universe 201

CHAPTER XVI.

Summary and Conclusion 205

EVOLUTION AND NATURAL THEOLOGY.

CHAPTER I.

ANCIENT AND MODERN VIEWS OF NATURE.

"And the Lord came down to see the city and the tower which the children of men builded."—Gen. xi. 5.
"Óðinn ok Frigg sátu i Hliðskjálfu, ok sá um heima alla."
GRIMNISMÁL.

IT is difficult in the present advanced state of physical science, to understand, or at least to realise, the aspect which the Universe presented to most of our predecessors, and which it presents even now to the uncivilised nations of the world, or to the uneducated among ourselves. The progress of science has been so great since the time of Galileo, that fresh vistas of knowledge have been opened up every few years, showing us a vast extent of country where we had been accustomed to see only a blank wall, with nothing beyond. First came the discovery of the shape, size, and subordinate position of the earth in the Solar System; then the boundless extent of the stellar universe was disclosed; then came Geology, with its disclosures of the vast age of the earth, and, later, of the comparatively great antiquity of the human race; then Chemistry, with its revelations of the common structure of the organic and inorganic world; and finally Biology, demonstrating the common origin of all life. The physical science of the ancient world never appears to have become the familiar heritage of the multitude, and what their wise men knew of the real system of the Universe, was lost or misapprehended after the decay of the ancient civilisations. Many of our sciences, too, appear to be exclusively of modern growth. A few generations back, and the wisest man did not even dream of facts now known to the humblest school-child: what then could the community have known or guessed of the nature of the Universe? No wonder that with the Ancients, and even more during the Middle Ages, the earth was either the Universe, or at least the centre and object of all things. It was supposed to be a flat plain, varying in size in different systems; but in all cases distant countries were regarded as unearthly regions, abounding with fabulous creatures of every description. Nor is this surprising, when we remember the size and ferocity of the wild beasts of the Quaternary period, with which men were certainly contemporaneous, and that many large and fierce beasts, such as the lion and the urus, have become extinct even in Europe, only within a few centuries at furthest. Thus, the rukhs of Arabian fable are merely exaggerated accounts of the large birds that recently abounded in many of the islands visited by Arabian navigators.*

The earth was formerly believed to be circular, and encompassed by the circumambient ocean, beyond which, in many systems, a chain of high mountains† formed a vast ring, the termination of the Universe horizontally. The earth itself was supposed to be hollow, or rather to form the roof of a great cavern (Hades, Sheol, or Hell),‡ inhabited by the souls of the dead. The

---

* The Islands of Wák-wák, of which such wonderful stories are narrated by these writers, are the Aru Islands near New Guinea, and the human heads which grow there like fruit, and cry "Wák-wák" at sunrise and sunset, are the Birds of Paradise, which settle in flocks on the trees and utter this cry.

† *Káf* of the Arabians, *Jötunheim* of the Norsemen.

‡ The two former words are generally translated "hell," in the authorited version of the Bible, which is unfortunate, as the latter word, though once synonymous, now conveys a very different idea.

B 2

visible heavens formed a hollow sphere, on the inner side of which the sun, moon, and stars were fixed, and it revolved round the earth in a day and a night. Of the real extent of the Universe, or even of the shape and size of the earth, the ancients had as a rule, not the remotest conception, although they occasionally attributed a definite size to the earth, far exceeding its actual dimensions.

In or above this hollow sphere was the dwelling place of the gods, who were regarded as beings very little superior (and sometimes actually inferior, even in power) to men, with whom they had constant intercourse face to face, as related in Genesis, the Homeric Poems, and in other writings of the mythic-heroic age. The gods were not eternal. In some systems, as in the Scandinavian, they were actually mortal; and in many mythologies, they lived in perpetual fear of revolutions which would bring their reign to a close. The Scandinavian gods were nearly all predestined to be slain in battle at the end of the world; and in Grecian mythology we find Zeus, the successor of two fallen dynasties, obliged to be extremely cautious in his matrimonial alliances, lest some one or other of his numerous sons should prove stronger than himself, and dethrone him as he had dethroned his own father. The same idea reappears in Judaism, Christianity and Islam as the Fall of the Angels.

We learn from the Edda that Othin is the fortunate possessor of a throne from whence he can survey the whole earth. Although this is but a primitive idea of omniscience, it is nevertheless far superior to the ideas entertained by the ancient Hebrews regarding the Elohim,* who were always obliged to descend to the earth in person, in order to obtain correct information about terrestrial matters,† and who appear not to have known that Adam had transgressed their commands, until the latter betrayed himself by his own conscience-stricken words and actions.

In the Indian mythology, the gods are often represented as inferior to men, and are compelled to resort to extraordinary expedients to

---

* The finite nature and attributes of the God (or Gods?) of the Pentateuch, who was little superior to the gods worshipped by other nations at the same period, and who is represented as less just and merciful than his servant Moses, are familiarly known and admitted; but it must always be remembered that the later Jewish writers entertained far higher ideas of the national deity; and that some of the noblest sentiments of the Gospels may be paralleled in the "uninspired" Talmud.

† Gen. iii. 8; xi. 5, etc.

maintain their own supremacy. The incarnations of Vishnu are well-known instances; but perhaps the most striking illustration of the powerlessness of the Indian gods is contained in the wild legend of the Sacrifice of Serpents, in the first book of the Maha-Bharata, where we find Indra, the so-called "king of the gods," forced to flee for his life from the presence of a human king and his Brahmins, and to abandon to their vengeance a suppliant whom he had sworn to protect.*

Were it necessary to pursue the subject further, it would be easy to multiply instances from every mythology to illustrate the finite nature of the deities of uncivilised nations, and it is clearly not accidental, but a special arrangement essential to human progress, that all finite religions thus contain the elements of their own destruction, and are therefore certain to be cast off one by one as civilisation and intelligence advance, to be replaced by faiths

---

* Adi-Parva, verses 2121–2128. As this remarkable myth is little known in England, an abstract is given in note *A* at the end of this chapter, extracted from the French translation of the Maha-Bharata, in which it is related at great length. Another, by E. Arnold, will be found in the *Journal of the Asiatic Society of Great Britain and Ireland*, n.s. vol. 14, pp. 259 & 260 (1882).

embodying higher views of the Unseen. Yet religious ideas change slowly, and it is much to be regretted when men cling to an out-grown deity, whose morality is not above but below their own standard and whose actions are unjust, absurd, or physically impossible, in the sense of being opposed to the known laws of nature.

Compare with the old-world teachings which we have just been briefly examining, our modern views of Nature and God, which, though they must always remain very imperfect, are yet far in advance of those of the ancients. Heaven and Hell, as localised regions, have vanished from the physical universe. Not that the ideas themselves to which these abstract terms are applied, have lost an atom of their significance; but as they pertain to other states of existence, they do not fall within the sphere to which our modern scientific methods of observation limit themselves, and hence we can no longer point to any region of the physical universe as the locality of Heaven or Hell. Besides, a single Heaven or Hell is no longer either scientifically or morally credible; and, although generally taught by Christians, yet the few remarks which have been handed down to us as Christ's own teaching on the subject certainly imply that he believed in gradations of both rewards and punishments* (Matth. xi. 22; Luke xii. 47, 48, xix. 17-26; John xiv. 2). Nor can we for a moment rationally suppose that even the highest and noblest of our race can reach the presence of the Infinite itself at one bound, from a world so low as this. Rather let us acknowledge with the Buddhists, the much greater probability that ages of progress, and possibly a great variety of stages of existence, separate us from Nirvana.†

Swedenborg's assertion that the sun of this world corresponds to the black centre of the infernal world, like his counter statement, in which most mystics of all ages have concurred, that God himself is the sun of the Spiritual Universe,‡ or at least of the heavens, as opposed to the hells, is obviously allegorical; but more than one writer of this and the last century seriously put forward the monstrous proposition

---

* The New Testament doctrine of a future state is misrepresented in the English translation. The word "heavens" has always been rendered by the singular, and the word "Hades," which simply means the world of spirits, generally by "hell."

† See note *B* at the end of chapter.

‡ See note *C* at end of chapter.

that the sun is literally the seat of the orthodox hell, forgetting the gross absurdity of making hell the material source of all life on the earth, and probably throughout the solar system. It would be far more reasonable to believe, with some modern French philosophers, that the sun is in some sense or other the heaven of its system; but such speculations are far beyond the sphere of inquiry to which the science of the present day limits itself, though (apart from our own individual experiences) they are not unlikely to fall within the range of the sciences of a not very remote future.

On the other hand, the field of scientific investigation which is universally admitted to be legitimate has been expanded to infinity. The earth has been weighed as in a balance; its form and size are approximately known; its various regions are daily being more fully explored; and the mysteries of the dust and the nebula are alike inexhaustible to the man of science. So far from the earth being the principal object in the universe, we now know that it is one of the smallest, and probably (except temporarily to ourselves) one of the least important of all the worlds which fall under human observation, while the utter destruction of the entire solar system would leave a void, as compared with the universe, infinitely less than would be produced in the earth by the loss of a single grain of sand from the sea shore.

Again, as the ancients had no conception of space, so also had they none of time. The world was thought to have come into existence, either just before or just after the gods, according to whether the gods were believed to have sprung from the world, or the world from the gods. Nearly all the early religions point to a very recent creation of man by the gods, or to his descent from them within a few generations. Queen Victoria can trace back her ancestry till two or three converging lines meet in Woden (or Othin) within much less than two thousand years. The more philosophical mythologies, such as the Persian and Scandinavian, represented that the earth would be destroyed at some future period, and subsequently renewed. But the ancients had no conception of the real age of the earth, although geological time is as nothing compared with astronomical. Now, however, our ablest geologists reckon the period of man's existence on earth (though perhaps geologically speaking, comparatively recent) not by generations, but by tens of thousands, if not by hundreds of thousands of years.*

Our theological ideas have also been expanded with the growth of our science, though not in proportion, because the influence of science on religion has hitherto been negative rather than positive. Homer in a well-known passage (Il. v. 339-342) explains the difference in the nature of gods and men to consist chiefly in the presence of a fluid called "ichor," in the veins of the former, instead of blood, the result of more wholesome food. So far from this, our modern views of the Unseen have been growing wider and wider, until Herbert Spencer asserts, in his "First Principles," that the only possible ground of reconciliation between science and religion consists in the recognition of an unfathomable mystery underlying both. Without going quite so far; it may perhaps be suggested that Science and Religion may at length meet on the broad platform of Theism and Philanthropy.

---

* Geology at present requires a longer time for the existence of the earth than astronomers are yet prepared to admit. But this discrepancy does not affect the certainty of the earth having been in existence for a vastly longer period than most of the ancients had any idea of.

No one who has ever paid the least attention to the vast changes in religious thought in England during the last twenty years, owing chiefly to the progress and wide dissemination of enlarged scientific truth (as true a revelation from God to the present age, as any religion has been to those which have preceded it) can doubt the great influence exerted by Science on Theology. Again, according to that law of instability, which is inseparable from the existing order of nature, all religious systems are continually undergoing disintegration and schism. Religion itself remains, for it is a part of the spiritual nature of man; but Theology, which is its varying mode of expression, changes from day to day; and it may now almost be said, that Materialism and Spiritualism* are fighting hand to hand over the prostrate forms of their common enemies, the orthodoxies of the past.

Religion and Science are the two great engines of human progress, each of which is deprived of half its legitimate influence if separated from the other, and we have only to wait for their

---

* I use this word here in its broad philosophical sense as the antithesis to materialism.

cordial union to behold all religious systems which are still hampered either with a cosmogony, or with any form of anthropomorphism, cast into the furnace, to come forth purified from finite ideas and more suited to the larger requirements of ourselves or our successors.

Note A.

THE SACRIFICE OF SERPENTS.

King Janamejaya once undertook a great sacrifice, to revenge himself upon the serpent Takshaka, who had killed his father. An immense fire was kindled, and the serpents, great and small, were compelled by powerful incantations to throw themselves into it. But Takshaka did not appear, and the officiating priests informed the king that he had fled to the court of Indra, who had promised to protect him. The king was enraged at this news, and commanded the high priest to redouble his exertions. He therefore sacrificed with irresistible incantations, and Indra himself appeared in a celestial chariot, surrounded by all the gods, who were singing his praises, and followed by troops of attendants. Takshaka, lying on Indra's robe, and trembling with fear, was also compelled to accompany him. Then the enraged king said to the priests: "Brahmins, if the serpent in Indra's chariot is Takshaka, hurl them both into the fire together!" The high priest then devoted Takshaka to the fire, and immediately Takshaka and Indra were seen to writhe about in great agony At the sight of the sacrifice, Indra was seized with terror; and abandoning Takshaka, fled back trembling to his palace. But just as Takshaka was about to be cast into the fire, the sacrifice was stopped at the request of a young Brahmin, to whom Janamejaya had previously promised a boon.

Note B.

BUDDHISM.

The meaning of the word Nirvana has been much disputed. Though differently explained by different sects of Buddhists, it is highly improbable that it usually means annihilation; it more probably denotes a state so far transcending anything we can even conceive of at present, that it can only be characterised by negations. Nirvana, in this latter sense, appears to be referred to in several passages of the New Testament (Matt. xxii. 30; Mark x. 30, etc.), which is not surprising when we consider the remarkable parallel which exists between Buddhism and Christianity, both as regards the characters of their founders, the essential principles of their morality, and the modern rites and ceremonies with which they have become encrusted at Lhasa and Rome. Buddha, however, was more of an ascetic than Christ, owing, perhaps, to his original royal birth having given him a greater contempt for the vanities of the world.

Note C.

GOD AS THE SPIRITUAL SUN.

All nations have worshipped the supreme Deity under the symbol of the Sun or Light; from the Persians and Jews (the latter of whom had their Burning Bush and their Shekinah) to the Christians, who represent a "glory" round the heads of God and the saints. But apart from the question as to whether the "glory" may not signifiy simply odic light, the Sun is still, with the possible exception of Alcyone, the grandest physical symbol of the Deity which modern science can present to us. And sublime as is the Hindu conception of the Trimurti with its three-fold attributes of the creating, upholding, and destroying powers of the Universe united in one, the symbol may also hide a physical mystery, and point to the triple powers of the sun's rays—the heat, the light, and chemical rays. The constitution of sunlight may have been known to the wise men of the east; and it would be worth while to enquire whether any hint of such a meaning is concealed in the word AUM, or any other of the sacred names of the Deity.

CHAPTER II.

THEORY OF DIRECT CREATION.

> "There grew an old Oak in the Vale of Elul,
> Old as the world, and planted in the Day,
> In that mysterious day, wherein God made
> The earth, and heavens, and each plant of the field,
> Before it was in the earth, and every herb
> Before it grew, while man as yet was not."
> HERAUD, *Judgment of the Flood*, v. 307–312.

SOME may think that we have already devoted too much space to subjects foreign to the Theory of Evolution; but it is impossible to insist too strongly upon the difference between the Universe as it appeared to the ancients, and as it appears to the moderns. We are always liable to overlook this difference, and consequently to misjudge both the ancients and their writings. On the other hand, it would be folly to undervalue or ridicule them on this account, for the highest genius is, and always must be of great rarity. It is not often that a man appears who is enabled by the mere force of his natural endowments to leave an indelible mark upon the ages for all time. Hundreds or even thousands of years may elapse between a Buddha and a Christ, a Homer and a Shakspeare, a Phidias and a Michael Angelo, or an Aristotle and a Newton. And while generations bow down in wonder at the unapproachable greatness of these masters of Religion, Art, and Science, some of whom have actually been deified by their admirers, yet such gigantic abilities as theirs are so rare as to be practically far beyond our powers of imitation. The mind, like the body, has its limits of possible development, varying in every individual, and in every age; and it would be a mistake to suppose that a man's mental and bodily capabilities depend entirely on his own efforts; for though he is very liable to underrate or overrate them, it is physically impossible for him to exert himself beyond them. Any really great genius must strike out a perfectly independent course for himself from the beginning, and advance in it alone. Competitors he may have; masters or rivals in his own age he can have none. He cannot begin where his predecessors left off; for to be their equal, he must stand as far above his own age as they stood above theirs. The influence of men of genius has raised the average standard of mankind; but it has not raised the masses to their elevation, and even were this possible, we may believe that the exceptionally great minds who would then arise amongst us, would be proportionably in advance of the masses. But with science, by which we may understand chiefly our knowledge of Nature and natural phenomena, the case is somewhat different. Although the influence of men of great original genius is occasionally necessary to give science an impetus in a fresh direction, or to enforce the acceptance of new truths, which in science as in other things, are always unpopular, yet men of ordinary abilities can carry the work forward by concentrating their studies within a sufficiently limited compass, and taking up the work of their predecessors where they had laid it down.

Physical science is, however, of modern growth, and was comparatively little studied by the ancients, although it is probable that they were far better acquainted than ourselves with certain branches of science more suited to the spirit of that day than to this.* Never-

---

* To estimate the ancient *psychological* sciences by the modern *physical* sciences, is to attempt to square the circle, or to add together ounces and inches. The mistake of ancient science lay in undervaluing Physics; that of modern science lies in undervaluing, or rather in gnoring Psychology. The reason for this will be explained presently.

C

theless, it would be a great mistake to attach much weight to any views bearing on *physical* science which have been handed down from ancient times, especially if connected with any of the unavoidably narrow religions then prevalent, of which, moreover, the *exoteric* side only, as a rule, has been preserved; the symbols, without the underlying truths. It is not intended to imply that great minds among the ancients were incapable of appreciating and discovering scientific truth, but rather that the spirit of the age was opposed to physical science, which was then empirical rather than experimental; and as the means of diffusing knowledge were also very imperfect, science could not then become cumulative as it now is, except to a very small extent, and within very narrow limits. Consequently, the general comprehension and diffusion of even the little which was certainly known, was impossible, and as a large amount of ancient learning was kept secret, we cannot wonder at its being almost completely lost during the wars and revolutions which accompanied the downfall of the Roman Empire.

The first teachers of Christianity naturally

and unconsciously adopted most of the scientific errors of their time, and when we find that not only the Apostles, but Jesus himself, never seem to have called in question the characters of the Jewish national heroes (many of whom were very bad men, and others purely mythological characters,*) but preferred to attack the vices of their own time, rather than those of the past, we ought not to feel surprised at their tacit acceptance of the Hebrew cosmogony, which it was still less in accordance with their mission to correct, even if we assume, in direct contradiction to the repeated assertions of Jesus,† that he and his followers were miraculously exempt from errors of every kind.

In early times, when men had sufficiently advanced in religious thought to connect the origin of the world with their gods, the next question which would arise would be, "How did the gods make the world?" What more

---

* Where this is not immediately obvious in the legends themselves, it is discernible from other independent and legitimate sources. Thus, on the strength of particulars given in some Hebrew commentaries respecting Noah, some German critics have regarded him as a solar hero.

† Mark x. 18, xiii. 32

natural than the supposition that it was either called into existence out of nothing, or that it was moulded out of previously existing materials? Accordingly we find that one or the other opinion is taught by almost every religion of which we have any knowledge. Neither of these views imply Evolution, though the second approaches nearest to it; but in view of the opposition encountered by Astronomy and Geology, and indeed every new science supposed, whether erroneously or not, to trench upon received theological opinions, it would certainly not be a presumption in favour of Evolution that it was accepted as a truth by the ancients, but rather the contrary.* They had not the least conception of geology and the allied sciences, without which the very idea of Evolution could not exist. Besides, our modern popular notions of religious cosmogony are not the legitimate outgrowth of the positive sciences of the more advanced portions of the ancient world, but the offspring of the ignorance and fanaticism of the Dark Ages.

Until the close of the last century, the theory

---

* Compare Herbert Spencer's "Biology," pt. 3, ch. 2.

of Direct Creation was practically unassailable. It was universally believed that the world was created about 6,000 years ago, and would sooner or later be destroyed. This rendered necessary but a single act of Creation, which is philosophically somewhat less incredible than a succession of such acts, as these would imply the direct, but occasional and capricious interference of the Almighty. But we now know that neither worlds nor individuals arise by direct miraculous agency without the intervention of secondary causes, and hence we are not justified in assuming this to be the case with species. Our conclusions must be guided by evidence alone. As far as our present knowledge extends, Matter and Force are fixed quantities; nor are we in any position to affirm that a soul is newly created when it is born into the world.

The nature of miracles is closely connected with the theory of Direct Creation. They were formerly supposed to be effects occurring independently of any appropriate cause, at the will of some superior being. In its more modern sense, a miracle becomes the action of laws which are either unknown or imperfectly understood, or perhaps of intelligent beings able to avail themselves of laws which are, under ordinary circumstances, more or less beyond human control.

The antecedent probability of an alleged miracle must depend partly on its scientific credibility, and partly on the evidence in its favour. In the Apocryphal Gospels,* Jesus is represented as making birds of clay, and then giving them life. Such a miracle (though analogous to the account of the creation of Adam, in Genesis, if taken literally) is obviously incapable of any rational explanation; but the New Testament miracles are usually of quite a different class.† Some few, such as that of the loaves and fishes are very difficult to explain; but even in this case, the incredible part of the miracle is confined to the *modus operandi*. Several miracles, identical in kind with that of

---

† On the whole, the divines to whose lot it fell to fix the canon of the New Testament, appear to have made a very judicious selection. Anyone who looks into the Apocryphal Gospels, cannot fail to be struck with their great inferiority, both in style, contents, and appearance of historical probability to the Four Evangelists. It is something like comparing the Sermon on the Mount with the Athanasian Creed.

† All the instances of raising the dead in the Old and New Testaments (except the Resurrection of Christ which seems to have been a case of long-sustained materialisation) appear, taking the accounts as they stand, to have been cases of suspended animation, or at most, of long continued trance. It is strange that Paul's recovery after stoning (Acts xiv. 19, 20) has not also been magnified into a miracle.

the loaves and fishes, are recorded of Elijah and Elisha, but without sufficient data to enable us to judge of the real agencies at work to produce the recorded effects, even if these accounts may be trusted. I am not aware of any precisely similar occurences recorded on good authority in modern times; but in principle they may be paralleled by that familiar class of special providences of which Müller's Orphan Asylum is a good example; but to the reality of which thousands of earnest men, belonging to every religious faith in the world, will bear grateful testimony.

If all authenticated miracles occur in accordance with natural law, the moral character of a miracle would cease to be an argument for or against the possibility of its occurrence, although there may be a deep truth in Bulwer Lytton's suggestion[*] that the highest powers of Nature are beyond the reach of any but the good. Neither Christ nor Mohammad relied upon miracles to prove their divine mission, but appealed to the character of their teachings. The first of these great prophets claimed, and apparently

---

[*] "Zanoni," b. iv. ch. ii.; Compare also R. Dale Owen's "Debatable Land," p. 121.

possessed the power of working miracles, but when he appealed to them at all, it was not to the power displayed in them, but to their beneficent character. Mohammad modestly disclaimed the possession of any such power,—no small argument to unprejudiced minds, that he was neither an impostor nor a fanatic.*

All the laws of Nature which can be studied at a distance (as connected with light, gravitation, chemistry, etc.,) are invariably found to be the same throughout the universe; and it is no unreasonable supposition that they are immutably impressed upon matter by the Almighty, and absolutely inseparable from its essence. Their universality is one of the strongest evidences which we possess for the unity of the Supreme Mind of the Universe, though this does not preclude the operation of one law upon another, with or without the action of lower intelligences, to produce infinite variations in their results; while the constant presence and supervision of the Supreme may well be imagined necessary both to the existence of the Universe itself, and to the maintenance of

---

* See also note to chap. xiv.

the unvarying action and proper equilibrium of the laws of Nature throughout the Cosmos.

As an anonymous writer has well remarked,* " Undisturbed harmony presupposes a power of maintaining harmony. We look out upon the heavens and behold a harmony and order absolutely intact. There is therefore somewhere the power of preserving, and there can be no power of disturbing that harmony, for else such influence would be manifest by desperate irruptions of disorder. We feel, we know that the heavens constitute a mighty system of order, and we do wrong to our native perceptions of analogy if we allow that there can co-exist with such a system, another whose motive is disorder, yet which does not show itself in disturbance on the other."

The unity of the system of Nature is a totally different thing from the unity of design in the construction of organic forms. The latter theory, as commonly interpreted by the doctrine of Special Creation, involves us in innumerable difficulties and absurdities. The theory of Prochronism,† which represents the world as created

---

* " Freelight," vol. i. p. 365). † Gosse's " Omphalos."

at a definite epoch, with fossils *in situ*, and bearing every mark of having existed from eternity, is, in spite of its obvious absurdity, the only view which can even partially reconcile the doctrine of Special Creation with existing facts. Nor is even Prochronism, although based upon Special Creation, necessarily opposed to Evolution, as this theory affirms that the facts of Geology are to be regarded simply as phenomena; not as representing anything that ever actually existed, but only as showing what *would* have existed, if the earth itself had existed for ever, instead of having only been created 6,000 years ago.

Wallace, in his well-known work on Miracles and Modern Spiritualism,* has pointed out that the present decline of the belief in the Supernatural is unprecedented in human history. It has been caused partly by the progress of science, and partly by the re-action against the extreme credulity of the Middle Ages, and can only be corrected by wider knowledge, showing us that the natural and the so-called supernatural, are both subject, under the control of the

---

* In a note on page 22. he adopts the present writer's views regarding the Witchmania and its results.

Divine Author of Nature, to equally immutable laws.

There can be little doubt that the real phenomena under-lying so-called miracles, witchcraft, mesmerism, spiritualism, etc., depend on ill-understood physico-psychological conditions, which are not equally developed in all persons.

In modern Europe alone has the crusade against all phenomena of alleged supernatural origin been waged with unremitting and exterminating fury; and in Modern Europe alone, have so-called supernatural phenomena become so rare as to lead to general disbelief in their reality; a result due to the operation of the ordinary principles of Natural Selection. During the Witchmania, all constitutions capable of producing even the simplest of such phenomena were weeded out, while those who possessed no such powers were left to continue the race. Moreover, such a persecution would be intensified by the higher classes of phenomena being the first to disappear, leaving only the lower forms to awaken more terror and aversion, till they also were quenched in blood. The result was somewhat analogous to that pointed out by Galton as produced by mediæval religious persecutions and celibacy.* The conditions under which these phenomena could be manifested at all, became rapidly rarer and rarer, and at last the phenomena themselves disappeared, being very nearly stamped out. Religious persecution spreads a belief, but constitutional peculiarities must necessarily be destroyed by such persecutions. Simultaneously with this disappearance of supposed supernatural phenomena came the vast revelations of modern science. Can we wonder that men whose immediate predecessors had almost succeeded in blotting out so-called supernatural phenomena, which they themselves had consequently no opportunity of personally investigating, should refer every story of the supernatural handed down from former times, to delusion, imposture, or to ill-understood philosophical experiments?

This idea would be increased by the great number of cases of alleged witchcraft which were recognised as based on nothing but imposture or suspicion, especially towards the close of the persecution On the principles of Darwinism, the witch-mania and the con-

---

* This subject will be further discussed in Chapter xiv,

sequent decline of belief in the supernatural in Modern Europe, stand to each other in the direct relation of cause and effect.

Jesus asserted that "prayer and fasting" were necessary to the performance of certain miracles, and cures are still effected by Eastern devotees, and at times by European Catholics by such means. Protestants do not fast; does this partially explain the absence of such cures among them? It has also been asserted that magic has been secretly practised down to the present day in many parts of Europe. All this, if true, merely illustrates our present argument, *viz.*—that the "supernatural" is as much the outcome of natural law, as the "natural."

CHAPTER III.

EVOLUTION IN ASTRONOMY AND PHILOLOGY.

> " Die Sonne tönt nach alter Weise,
> In Brudersphären Wettgesang,
> Und ihre vorgeschriebne Reise
> Vollendet sie mit Donnergang."
>
> GOETHE's *Faust*.

BEFORE discussing the bearings of the Theory of Evolution on Biology, it may be well to examine its connection with other sciences, in order to ascertain whether they present any traces of its operation, or any analogies which will justify us in reasoning from them to Biology. We will therefore consider Astronomy and Philology, two sciences as little connected with each other as possible, and consequently very well suited for our present purpose. Geology is more immediately connected with Biology, and will therefore require to be dealt with subsequently.

Astronomy reveals to us the existence of matter in all stages of development, from the vast masses of pure hydrogen, of which some nebulæ appear to be composed, to the comparatively solid orbs, varying infinitely in density and composition, known as suns and planets. Nothing in the universe is stationary; everything is changing place and state at each moment of its existence. It is only true in a very general sense that "the thing that hath been, it is that which shall be, and that which is done, is that which shall be done, and there is no new thing under the sun."* If we reflect a moment we perceive that everything which is, must depend upon everything which has previously existed in the eternal past, and that what has once happened, can never happen again under precisely the same circumstances, unless the plan of the universe is that of an endless succession of cycles, exactly repeating each other, an Oriental speculation which probably finds few supporters among ourselves. But for the slight variations in our surroundings, which we do not notice because we are so used to them, life would perhaps soon become too monotonous to be borne.

The periods with which Astronomy deals are so vast that the higher cosmical phenomena

* Eccles. i. 9,

cannot often be expected to fall under human observation at all. It is however, tolerably well established that the various worlds which are scattered through space, run through a certain course of development, and perhaps also of destruction. The laws of Energy clearly indicate that the universe is not eternally stable, and that it cannot exist for ever in its present form. But all analogy, even if scientific data are at present wanting, will justify us in assuming that its dissolution would be merely a prelude to a yet grander future development. The history of the earth, as a whole, furnishes us with no instance of a retrograde movement; and our knowledge of its past is necessarily incomparably greater than of that of any other world.

Herbert Spencer believes that the Universe, in whole or in part, is continually passing from a nebular or an amorphous, to an organised state, and *vice versâ*. But we are not justified in believing that these changes are "Kalpas," during which, every event of previous cycles repeats itself over and over again; and all our knowledge is opposed to a speculation, which would reduce the System of the Universe to a recurring decimal.

It has been supposed by some writers * that the planets are receding from the sun, and will ultimately become suns themselves; but it is a more general opinion, and more in accordance with received scientific theories, that the reverse is the case, and that the planets are gradually approaching the sun, and will ultimately become absorbed into its substance.

A very remarkable phenomenon was observed in the star T Coronæ Borealis a few years ago, which appeared to indicate that a planet possessing an ocean had fallen into its primary.†

If therefore, we accept the teachings of the Mystics, and look upon God as the Spiritual Sun, analogy would at once lead us to the doctrines of Emanation and Absorption, which form so important a part of the profound philosophies of the East, and are not regarded with disfavour by many unprejudiced thinkers among ourselves.

Little is at present known of the nature of the more distant planets; but the largest are those furthest from the sun, and are also those

---

* Luke Burke, and J. W. Jackson, for example.
† See *Good Words* for April, 1867. A similar appearance was presented by another star in November, 1876.

most resembling it in their physical characteristics. The belt of asteroids, which some astronomers suppose to be the fragments of a shattered planet, forms the boundary between the smaller and larger planets. It is not thought unlikely that some of the latter (Jupiter, for example) are capable of radiating a sufficient amount of light and heat to maintain life on their satellites. This appears the more probable when we remember that it is now known that even the moon reflects a perceptible quantity of heat, as well as light to the earth.*

It has sometimes been argued,† but on very insufficient grounds, that the earth is the only planet in the Solar System, if not in the Universe, which is physically capable of supporting life ; but, apart from the inherent improbability of this theory,‡ the amount of light and heat required by each planet for the use of its inhabitants, is probably dependent as much on the constitution of its atmosphere, and the amount of its own internal heat, as on its actual distance from the sun. Astronomers have supposed that

---

* Piazzi Smith, "Teneriffe," p. 212.
† Whewell's "Plurality of Worlds."
‡ See Ponton, "The Beginning," ch. ix.

even the sun itself may be inhabited, and that its chemical activity is limited to the photosphere, which they imagine to be separated from the body of the sun by a screen of dense clouds. It was even at first suggested that the so-called "willow-leaves" of the photosphere might be living organisms of some description.

A short time ago, it was believed that the Galaxy was a nebula, or star-system, to which all the visible heavenly bodies (except the other nebulæ, then regarded as independent star-systems) belong, but spectrum analysis has now shown us that all nebulæ are not of the same constitution, and it is at present held that they are not star-systems totally independent of our own, but masses of matter in a less organised condition than the rest. According to this view, the whole of the visible heavens, whether stars or nebulæ, would form one vast system, though some portions, such as the Galaxy, the Magellanic Clouds, some of the resolvable nebulæ, and various star-clusters, appear to form subordinate systems of their own, more or less independent of the remainder.

Apart from all speculation, it is incontestable that the moon is more or less dependent on the earth; the earth on the sun; and even the sun on Alcyone (if astronomers are correct in believing that star to be the centre of the whole star-system), while it is not unreasonable to suppose that this system itself is but one of an innumerable multitude scattered throughout space. The Nebular Hypothesis is now the most commonly received theory of astronomy; and it supposes the stars to have been gradually condensed out of previously existing gaseous matter. Be this as it may, all cosmical changes are very gradual, and it is universally believed that every star has gone through, or is passing through a series of changes analogous to those of the earth. We are therefore fully justified in regarding Evolution as a recognised principle in Astronomy.

We will now turn to Philology, and enquire what is its teaching? We find in Europe a great number of languages and dialects, more or less related to each other, most of which could readily be classified, like natural objects, into families, genera, and species. Some of these run into one another, while others are separated by considerable gaps.

The history and literature of Europe, joined

to a study of the European languages themselves, leave little doubt as to the immediate origin of most of these tongues, and in some cases, we are even able to reconstruct portions of unwritten history from the impress it has left on our most familiar expressions.

But amongst savage nations, as for example, the aborigines of Africa, we sometimes find every tribe, or even every village speaking a different dialect, and language in so unstable a condition that it has sometimes happened that when the children of a tribe are left much to themselves, they grow up speaking a new language, unintelligible to their parents. Most savages have a very poor vocabulary, and make great use of gestures to explain their meaning; and it is possible that language may have originally grown out of a combination of ejaculations with gestures.

Another cause which renders savage dialects very fluctuating, is the dislike of many tribes to utter the names of the dead. This is carried to such an extent among the Abipones of South America, that when a man dies, they alter every word in their language which has any resemblance to his name.

In the two last paragraphs, we have been speaking of the instability of very poor languages. More copious ones, and those spoken by larger communities are much more stable. The increase of literature has a great tendency to keep the languages of modern Europe fixed; but the chief disturbing influence arises from the constant communication between natives of different countries now speaking widely different languages, which leads to the continual introduction of new words. But on the other hand, constant intercourse between tribes speaking a similar language. has a great tendency to permanency of dialect. Hence the resemblance between the Semitic languages, at the head of which stands Arabic, one of the most ancient, copious, and pliable languages in the world, which has been spoken for thousands of years over an immense extent of country; and which gave rise to the remark of Higgins ("Anacalypsis," i. p. 465) that, "In the character of *nomade tribes* there seems something favourable to the retention of language, which is difficult to account for." It is easier to comprehend now; for, applying to Philology the recognised principles of Zoology, we perceive at once that the constant intercrossing of closely allied dialects would naturally prevent the differentiation of new languages.

The languages of Europe may be considered to have reached a stage of development analogous to that of the higher animals, while those of savage nations may more fitly be compared with that of the very lowest forms of life, these masses of sarcode of no determinate shape or size, but constantly fluctuating in both.

Let us suppose that an intelligent being from some other planet, who was unacquainted with speech, and possessed some other mode of communicating his thoughts, were to pay a visit to Europe. His first impression might be that men were supernaturally endowed with the gift of speech, and that each separate language and dialect, no matter how close their resemblances, were communicated to the nations speaking them by direct divine revelation. But even without the study of comparative philology, a very slight knowledge of history would be sufficient to lead him to doubt the correctness of his first impressions, and to form some other theory more nearly approaching the truth. In like manner,

Herbert Spencer* represents an ephemeron as reasoning that men and women must have been specially created as adults, because he could see no change in them in the course of his own existence. Such, in sober truth, is the actual position occupied by naturalists during their investigations into the various problems connected with the Origin of Species.

---

* "Biology," pt. 3 ch. 2.

CHAPTER IV.

EARLIER THEORIES CONNECTED WITH EVOLUTION.

"Some say he bid his angels turn askance
The poles of earth, twice ten degrees and more
From the sun's axle."
     MILTON, *Par. Lost*, x. 668-670.

"He also deprived the serpent of speech, ... and when he had deprived him of the use of his feet, he made him to go rolling all along, and dragging himself upon the ground."
    WHISTON'S *Josephus, Jewish Antiquities*, ch. i.

For many centuries the science of Natural History made very little progress, and consisted of an assemblage of unconnected facts and descriptions, mixed with fables. At length in the middle of the last century, Linné laid the foundation of systematic Natural History, and thus rendered possible its study as a progressive branch of science. The great object which the earlier naturalists proposed to themselves, was the discovery of the true System of Nature. It was thought that organic life formed an unbroken chain, and that the ignorance of even one link would prevent our being able to form a correct idea of the whole. One of our poets has expressed this idea as follows:

> "Each moss,
> Each shell, each crawling insect holds a rank
> Important in the plan of Him who framed
> This scale of beings, holds a rank, which lost,
> Would break the chain, and leave a gap between
> That Nature's self would rue."

But this view was propounded in ignorance of the enormous number of species which have become extinct, the vast self-adjusting economy of the true System of Nature, and the most obvious facts of Geographical Distribution. Species were then numbered by fewer hundreds than we now reckon thousands, and every specimen the exact locality of which was unknown, was vaguely entered as " from the Indies ;" although a naturalist of any experience can now frequently make a tolerably accurate guess at the locality of any unlabelled specimen before him, much more of a series of specimens. As the knowledge of species increased, it was soon discovered that there were many side-links which would not harmonise with the idea of a connected chain, and that there were frequently great gaps between species, as if the intermediate links were wanting. This rendered the supposed chain very broken and irregular, and the idea of a single connected chain of organisms was gradually given up. The Circular System was then suggested as the true System of Nature. This hypothesis supposes that each three, five, or seven species, genera, etc., formed with a central type, a complete circle returning into itself, and touched on all sides by surrounding circles. This system, though a great advance upon the former method, also proved to be incomplete. The fallacy of the circular classifications founded on one or other of the supposed sacred numbers has frequently been exposed, and it is now clear that neither the linear nor the circular arrangement can be maintained, and that nature cannot be conformed to any such regular system of classification.

Some of the classical writers appear to have propounded theories of nature more or less allied to evolution,* but De Maillet in 1735 was perhaps the first author who promulgated an hypothesis similar to those now under discussion, relating to the Origin of Species. Some paragraphs in the Prolegomena to Linné's " Genera Plantarum" (ed. 6, 1764), have been brought

---

* Aristotle, "Physics," ch. 8.

forward to show that Linné himself had doubts as to the immutability of species; but the exact meaning of these passages is somewhat uncertain. Towards the close of the last century, Goethe, Geoffroy Saint-Hilaire, and Erasmus Darwin simultaneously began to express doubts respecting the correctness of the current opinions as to the immutability of species. Their views were subsequently stated more fully by Lamarck, and were popularised by Robert Chambers, and similar opinions have been advocated from time to time by later authors.

But the writings of the earlier Evolutionists were far from satisfactory, owing to the want of sufficient materials upon which to base a tenable scientific theory, which sometimes led them into vague speculations likely to create a prejudice against their views. Some thought that the combined action and volition of an animal would gradually develop limbs which it did not previously possess, as if any animal could conceive the idea of powers absolutely beyond its experience, and foreign to its nature.*

---

* Bree ("Exposition of Fallacies in the Hypothesis of Mr. Darwin," p. 211) erroneously credits Wallace with teaching similar absurdities.

Others supposed that new species might have been produced by the intercrossing of old ones; but this cause is now known to be physiologically almost impossible in a state of nature, though it appears to have played a part in the formation of the breeds of some of our domestic animals.

Another view, advocated by Buffon, who thought that the ass was simply a degraded variety of the horse, was, that new species may have arisen from degradation of character. His illustration was peculiarly unfortunate; for the wild ass is a creature as well adapted to its surroundings as the wild horse; and although congeneric, cannot be regarded in any sense as a degraded creature. Although degradation may possibly have occured in individual species, and others appear to retrograde in their metamorphoses, yet the great majority of existing species are certainly not inferior to their predecessors; and the continued differentiation between existing organisms seems to be sufficient to insure steady progress on the whole. Apparent retrogression, as exhibited in the extinction of species and individuals, will afterwards be shown to be the necessary ally of real progress.

Several mediæval and modern theologians have held the Degradation Theory in a sense similar to that alluded to by Milton, in the passage quoted at the head of this chapter. They maintained that a great physical revolution took place at the Fall, overturning the whole original system of Nature, and rendering necessary its re-organisation on a very inferior scale.* Whatever may be the true interpretation of the ancient legendary narrative in Genesis,† it is palpable that no such universal cataclysm has ever occurred, or it would have left ineffaceable geological traces.‡ Nor does the Biblical record itself allude to any cataclysm at the time of the Fall, but simply states that man, the serpent, and the ground were cursed in consequence. The legend itself must in any case have been handed down by human agency, and certainly cannot be interpreted literally, as even " the primal curse " meets with no support from Geology, which is God's handwriting on

---

* See Hugh Miller's " Testimony of the Rocks," Lecture x.
† On account of the importance often attributed to Genesis in these discussions, I have added a note on the subject to the present chapter.
‡ Formerly, geologists thought they saw in every dislocated series of strata, traces of Noah's flood; now the most pious can find no phenomena which they can attribute with certainty to its influence.

the rocks, and absolutely infallible in so far as our apprehension of it (always of course, limited by our knowledge) is correct; whereas, if we regard Inspiration as a communication received from higher beings, through intermediate agencies, it will always be limited by the imperfections of both the communicants and recipients, and must therefore be necessarily opposed to infallibility. It may well be doubted whether Absolute Truth could be communicated to us at all, or even if it could, whether it could possibly be made intelligible to us.

It is impossible to doubt any longer that men existed on the earth for thousands of years before the Biblical chronology, nor have we any reason to suppose that the race as a whole, has retrograded. Serpents, organised as at present, existed long before the date assigned to Adam; and although perhaps derived originally from four-legged creatures of some kind, cannot have been suddenly deprived of their legs (to say nothing of their power of speech, as asserted by Josephus, who certainly gives us the orthodox Jewish interpretation of the passage in his day); if only because every gradation between four-legged lizards and legless serpents may still be met with. We must therefore conclude that it is a mistake to interpret the earlier chapters of Genesis literally, especially as all attempts to reconcile them with modern science have hitherto proved equally unsatisfactory both to theologians, and to scientists, and have generally been shown, by fresh discoveries to be intenable, almost as soon as propounded. It is impossible to value too highly the infallible teachings of Science, which shews us the real light in which to regard ancient writings, that are said to be God's revelations to man in former times, but which are now seen to be not only unscientific but sometimes even immoral, in the clear light of God's truth, which we are privileged to enjoy. While retaining and admiring whatever they contain of permanent importance, we are bound to estimate at its real value everything which proves to be false or temporary. We must always remember that Truth must ultimately prevail in both Religion and Science, and that anything in either which shuns the light of a *fair* and thorough investigation, must be rejected as almost certainly intenable.

NOTE TO CHAPTER IV.
THE BOOK OF GENESIS.
(*See p.* 46.)

IT is certain that a large portion of Hebrew literature was lost at the time of the Captivity, subsequently to which the surviving portions were edited, and possibly enlarged or curtailed, by Ezra, who would draw on the memories of the elders of the race as well as on his own memory for missing portions.*

The book of Genesis consists of mingled tradition and allegory; and it may perhaps be doubted whether the story of the Creation is of Egyptian or of Chaldean origin. The early Christian writers, from Paul downwards, as well as Philo, treated the Old Testament as allegorical; but although this alone should teach us that we are not bound to accept everything literally, it is more than probable that the real esoteric meaning of the allegorical portions was unknown to them; and it is often concealed (either intentionally or through ignorance) in the Septuagint, as when the word *Elohim* is translated in Greek as in English, simply Θεὺς, God.

If Higgins is correct in his ingenious interpretation of the esoteric meaning of the first verse of Genesis, "By Wisdom the Trimurti regenerated the planets and the earth," the narrative of the Creation embodies the profoundest speculations of Oriental philosophy, perhaps mingled with traces of primeval science. The word translated "created" has no such strong meaning either in Hebrew or Greek, but simply means "to found, to make, or to construct." It will readily be seen that my remarks in this work respecting the Old Testament apply only to the exoteric or conventional interpretations of its contents, with which we are all familiar.

---

* 2 Esdras, ch. xiv.

CHAPTER V.

DARWIN AND HIS CRITICS.

"Flower in the crannied wall,
. . . . . if I could understand
What you are, root and all, and all in all,
I should know what God and man is."
                                    TENNYSON.

"By properly understanding *one atom* with all its constituents, the mind would be comprehending a globe."—A. J. DAVIS. *Principles of Nature*, p. 154.

DURING the course of the present century, various authors have written with more or less ability in favour of the Origin of Species by gradual Evolution, but the theory gained little support either from naturalists or from the general public, until the simultaneous promulgation of the Theory of Natural Selection, by Darwin and Wallace, inaugurated what may well be termed a new era in philosophical natural history.

A very compact view of this theory has been given by Wallace,* and also by Mivart,† and it

---
* "Contributions to the Theory of Natural Selection," p. 302.
† "Genesis of Species," p. 5.

will be convenient here also to sum it up in as few explanatory words as possible, though somewhat more fully than in the passages just quoted.

The calculations of Malthus and subsequent writers, show that all living organisms tend to increase in a geometrical ratio; and therefore, any single species of animal or plant (however small, or however slowly it might reproduce its kind) if allowed to multiply unchecked, would speedily overrun the earth.

But instead of this being the case, the actual number of individuals of any given species existing at one time, remains almost stationary; and hence the number of deaths must on an average equal that of the births.* It therefore follows that every species is exposed to adverse influences which keep its numbers down to about the average, and thus prevent any inordinate increase.

Every individual of a species resembles its parents in most particulars, but is subject to an indefinite number of slight modifications, which

---

* Greg, in his essay, "Malthus Notwithstanding," argues that other checks besides death may sometimes operate in preventing an undue increase of population. See his "Enigmas of Life."

are liable to be transmitted to its progeny, and perhaps intensified. If any particular modification conferred an advantage on its possessor in its reactions on other species, or on other individuals of the same species, which would have a tendency to prolong its life, the new variety would be more likely to be perpetuated than the old one, while the original form of the species, not having equal advantages, would have a greater tendency than before to become extinct. These slight modifications might be useful in a variety of ways, as for example, by enabling their possessors to resist changes or extremes of climate, to procure food more easily, to conceal themselves better from their enemies, to multiply more rapidly, etc.

There appears to have been ample time since the origin of life on the earth for the origination of all organic forms from some primeval form of very low position. Sir W. Thomson, indeed, calculates* on astronomical data, that life cannot have existed on the earth longer than about 100,000,000 years. But this calculation is rendered doubtful by the consideration that

---

* "Trans. Geol. Soc. Glasgow," vol. iii.

the conditions of life upon a planet must needs be regulated by the nature of its atmosphere and the amount of its internal heat, as well as by its distance from the sun, and the actual amount of imponderables received from this source. In the same way, it is inconsequential to argue that life cannot exist in planets nearer or further from the sun than our own; for it is quite conceivable that each planet may absorb a sufficient amount of imponderables from the sun for its own necessities, whatever its distance. Such an argument could only hold good if the planets were all of similar constitution to the earth; but although probably composed of the same materials, we know that they differ in size, density, and probably in the composition of their atmospheres.

Sir J. Lubbock* argues, that while geological data are still very doubtful, they indicate a very much greater antiquity for the earth than that assigned to it by Thomson. The denudation of the Weald alone is estimated to have required more than 150,000,000 years. Wallace

---

* "Prehistoric Times," ch. xii.

suggests that species may have been exceptionably stable for the last 60,000 years or so.

Many objections have been made to Darwinism, some utterly futile, and others demanding the gravest attention. Various writers are opposed to Biological Evolution altogether, while others, though accepting Evolution, do not think that the theory of Natural Selection is sufficient, alone, to cover all the facts. Some of these objections we will notice here, beginning with the theological argument that Evolution is opposed to the Bible.

Anything which is opposed to the previous convictions of the multitude, whether directly connected with Religion or not, is nearly sure to be met with this outcry, which is essentially the same as that encountered by Astronomy and Geology when they were passing from the stage of speculative or empirical to that of deductive sciences. When the science of Biology arrived at the same stage it could not expect to escape the same opposition.

Dr. Bree has published an "Exposition of Fallacies in the Hypothesis of Mr. Darwin," in which he claims a right for the "scientific believer" to challenge scientific theories which conflict with his opinions. But no such right can be conceded (on *a priori* grounds), even to an educated Protestant, for this, I presume is Dr. Bree's meaning by a scientific believer, without its being also granted to those holding other forms of religious belief, and as every system of religious cosmogony contains both scientific fallacies, and is inconsistent with every other, it is manifest that although every man has a right to ask for proofs of any statement opposed to his previous opinions, yet it would be a great obstacle to the progress of knowledge if any man's previous opinions were allowed to form a presumption against the correctness of any new scientific theory or discovery.

Dr. Bree seems to have misunderstood the meaning of scientific hypothesis; which is not a settled formula like a theological creed, but a supposition which may or may not be true, but which, by co-ordinating a number of facts which do not admit of a more plausible explanation in the present state of our knowledge, is of great temporary service to the progress of science, until it is either confirmed or superseded by enlarged knowledge. The real obligation of Mr. Darwin's opponents is not to show that this or that portion of his views is untenable, but if they reject his theory, to substitute another which will furnish a better explanation of the facts which he attempted to explain by it.

Dr. Bree also assumes (p. 60) that there must have been a beginning. How does he know this? He is speaking of absolute beginning, not relative beginning, and has overlooked the difference between the two. Again he says, "the special creation hypothesis is simply a belief." So is the hypothesis of the creation of the world in six days, which no scientific man can any longer profess to believe literally. The theories of Evolution and Special Creation are simply theories of the mode of operation of God's work in the universe and by no means involve the alternatives of Theism or Atheism. Which theory is true, must ultimately be decided by facts, and which is most consistent with God's general mode of operation, by observation and analogy.

We may even ask how a man who speaks of the necessity of a beginning, can logically believe in the eternity of God. Is it not conceivable that the universe may even be a necessity to God, as God is a necessity to the Universe?—the one eternal and ever-changing under the superintendence of its eternal and unchangeable Lord. Even yet the finite ideas of the ages of mediæval ignorance prevent our realising the vast revelations of Himself which God has vouchsafed to us by the light of modern science.

Yet the Darwinian hypothesis is far less destructive of the lingering anthropomorphism of the popular theology than the now universally accepted facts of Astronomy. Without enlarging further on this subject, it must be admitted that men of science frequently exhibit the same kind of incredulity in matters opposed to their established convictions, as theologians. Not to speak of still more recent instances, Lubbock [*] remarks that the first account of human remains having been found in company with those of the great extinct mammalia, was refused publication by the Geological Society as being too incredible for belief; but it was merely the first statement of what is now an established fact. All honour to men like Galileo, Darwin, Wallace

---

[*] "Prehistoric Times," p. 306.

and Lyell, who dare to stand up and proclaim what appears to them to be the truth, no matter how it may conflict with the prejudices of their age and country.

Another series of objections commonly brought against Darwinism, relate to the origin of man. Some of these arise from misconceptions, such as when Darwin is supposed to affirm that men are descended from existing animals. But Darwin certainly did not assert that men are descended from any animals now in existence, but merely that existing animals give us an analogical clue to the probable line of the specific descent of man, which is a very different matter.

Wallace regards Natural Selection as wholly insufficient to account for the origin of man, while E. Newman claimed that man was originally made in the image of God, and has continually degenerated, instead of improving. These points will, however, be discussed in a subsequent chapter.

An argument upon which some writers lay great stress, is the existence of beauty and design in Creation, which they conceive can only be explained by the direct action of the Almighty. No one will deny that beauty and far-reaching design beyond all that has ever been imagined by man exists throughout Nature; and Natural Selection shows us how some at least of these effects have originated by the action of laws within our comprehension; but this does not prove that they were not foreseen and fore-ordained by the Creator, who if omniscient, must have been aware of the entire results of his work, whether performed by Special Creation or Evolution.

A series of objections to Natural Selection have been stated by Mivart,* which it will be well to discuss here at some length, as they are of considerable importance :

> I. " Natural Selection is incompetent to account for the incipient stages of useful structures."

Murphy's argument that the eye could not have been developed by Natural Selection is one illustration selected. But animals are believed to have arisen from those very low forms of life which are all skin, all stomach, all limbs, etc., as occasion requires; and it is probable that

---

* "Genesis of Species," p. 21. These arguments are fully considered and replied to by Darwin, " Origin of Species," 6th ed. ch. vii.

they are also all eye, in the sense that all parts of their body are sensitive to the influence of light. That the sense of sight should in the course of time have been located in special organs which would require to be adapted for the purpose, and would be subject to Natural Selection, is scarcely more surprising than that limbs have taken the place of pseudopodia in all animals except the very lowest.

> II. "Natural Selection does not harmonise with the co-existence of closely similar structures of diverse origin."

As Mivart admits that this objection only applies to the Origin of Species by Natural Selection *alone*, it is unnecessary to consider it here, our subject being not Natural Selection *per se*, but Evolution.

> III. "There are grounds for thinking that specific differences may be developed suddenly instead of gradually."

Two instances will suffice as illustrations. Oyster spat taken from the British coast to the Mediterranean, produces the spiny Mediterranean oyster *at once.** And on several different occasions, the black-shouldered peacock has appeared suddenly in different flocks in England, sometimes increasing so rapidly as completely to supplant the common kind.† The black-shouldered peacock is unknown in a wild state, yet it is so unlike the common form, that some of our most eminent ornithologists have not hesitated to regard it as a distinct species, reappearing in consequence of the breed having been formerly crossed with it. But most naturalists are now of opinion that this form is no more an independent species than any of the breeds of our domestic animals.

Andrew Murray‡ was also of opinion that species might arise by sudden variation. He considered that a species might remain tolerably stable during a long period, from the absence of any special stimulus to variation, till an impetus was given to it by some change in surrounding conditions, and that a species

---

* Costa, " Bull. de la Soc. Imp. d' Acclimatisation," tome viii. p. 351, quoted by Darwin.

† Darwin, " Variation of Animals and Plants under Domestication," vol. i. p. 291.

‡ " Geographical Distribution of Mammalia," ch. ii.

which has once received such an impetus is liable to vary more and more, till it again settles into a state of comparative stability. He instanced the imperceptible formation of the Europeo-American race as an illustration of the manner in which new species or races may arise before our very eyes, without our being able to distinguish the steps by which the modification was brought about.*

It can scarcely be doubted that both gradual and sudden variations have acted in the production of the system of nature as we now see it. In cases analogous to that of the domestic pigeon, where there is *ab initio* a strong tendency to variation, it is probable that gradual modification may have sufficed in time for the differentiation of species; but in cases similar to that of the peacock, a bird which varies very little under ordinary circumstances, it is more likely that variation, if it occurs at all, will occur suddenly, like a " sport " among plants.

---

* Our common Small White Butterfly (*Pieris Rapæ*) has lately been introduced into Canada and the United States, where it has completely established itself. An extraordinary yellow variety almost unknown in Europe, is of frequent occurrence in America; and it is quite possible that this may ultimately become the ordinary American form.

IV. "The opinion that species have definite though very different limits to their variability, is still tenable."

This is fully admitted by Wallace,* who remarks, "In the matter of speed, a limit of a definite kind as regards land animals does exist in nature. All the swiftest animals, deer, hares, antelopes, foxes, lions, leopards, horses, zebras, and many others, have reached very nearly the same degree of speed. Although the swiftest of each must have been for ages preserved, and the slowest must have perished, we have no reason to believe there is any advance of speed. The possible limit under existing circumstances, and perhaps under possible terrestrial conditions, has been long ago reached." Murphy,† however, suggests that the limits which undoubtedly exist to unlimited variation in one direction, may perhaps be due to the counteracting influence of a tendency to reversion; and that if we could keep our racehorses (say) at their maximum of speed during a sufficient number of generations, this tendency might in time be

---

* "Natural Selection," pp. 291–294.
† "Habit and Intelligence," vol. i. p. 218.

bred out, and the race might again begin to vary in the direction of increased speed.

V. "Certain transitional forms are absent, which might have been expected to be present."

This objection assumes that the geological record is more perfect than was supposed by Darwin. Lubbock* points out (on Lyell's authority) that the Quaternary Mammalia are frequently intermediate between existing species. Thus, the European and American bisons are blended by remains meeting in *Bison Priscus*, and the brown and grizzly bears meet by imperceptible gradations in the cave bear; but no naturalist would class the existing bisons or bears as varieties, though their common origin must be regarded as proved. It would be highly interesting and important to ascertain whether those distinct species which are geologically one, are still capable of producing fertile offspring, if crossed.

But the Quaternary period is the geological age immediately preceeding our own; and yet how many species which then existed in

---

* "Prehistoric Times," ch. ix.

abundance, are only known to us by a few fragments of bone; and how many more must have utterly perished, especially among species of small size, without leaving even a vestige of their existence behind, although they may have numbered hundreds of thousands of individuals! When we consider the upheavals, depressions, denudations, floods, earthquakes, and volcanic eruptions which have acted on and metamorphosed the crust of the earth for countless ages, and then remember how small a portion of its surface has yet been geologically examined, we cease to wonder at the deficiency of connecting links, whether between individual species, or entire Classes, and have rather cause to be surprised that, under such unfavourable circumstances, we should already know even as much as we do of the past life of the earth.

VI. "Some facts of geographical distribution supplement other difficulties."

Geographical distribution will be discussed in a subsequent chapter, and need not be dealt with in this place.

VII. "The objection drawn from the physiological difference between 'species' and 'races' still exists unrefuted."

This is indeed the great stumbling-block to all theories of Evolution, nor has any perfectly satisfactory explanation yet been given of the sterility of most species when crossed, and the fertility of most varieties. Darwin * however, in his usual masterly manner, has brought forward a great number of facts and arguments to prove that this rule is not absolute, and varies very much in its operation. He thinks that the sterility of hybrids is due in many cases to the early death of the embryo; but this explanation is neither very clear, nor very satisfactory, and the following may be suggested as perhaps the true one.

It is well known that crossing promotes reversion, and that many plants and animals are normally hermaphrodite; but it is certain that hybrids frequently show more or less tendency to hermaphroditism, and if this effect is due to reversion, it will at once satisfactorily account for hybrids being frequently sterile, even where

---

* "Origin of Species," ch. viii.

no apparent reason exists. The sterile male hybrids between the pheasant and common fowl take great delight in sitting, whenever they can find a nest of eggs unoccupied by the hen.* A more important fact is that though hybrids can be obtained between two allied hawk-moths (*Smerinthus Ocellatus* and *S. Populi*), the sexes are nearly always mixed, as well as the species. The normal larvæ discharge either a white or a yellow fluid when touched, according to sex; but the hybrid larvæ discharge both.† Mr. Jenner Weir informs me that the mammæ are unusually large in male mules in Spain. As the sexes become very gradually separated as we ascend in the animate scale, it appears highly probable that reversion may play some part in this tendency to hermaphroditism in hybrids. It may here be suggested that some breeds of cattle, among which hermaphroditism is exceedingly frequent, may have sprung to a greater extent than others, from a mixture of several wild species.

Having now briefly glanced at some of the

---

* Darwin, "Animals and Plants under Domestication," vol. ii. p. 52.
† Trans. Ent. Soc. Lond. vol. iii, pp. 193—202.

more important objections to Evolution and to Natural Selection, we will proceed to examine a few of the principal facts which support these theories, and the various considerations which they involve. It must be clearly understood that Evolution does not necessarily imply Natural Selection; for although the latter is undoubtedly one of the principal laws which control the development of organized beings, it is equally evident that we should be pressing it too far to regard it as the key to *every* mystery in Nature; an importance which Darwin himself never claimed for it, but on the contrary, emphatically repudiated.

CHAPTER VI.

USE AND DISUSE, REVERSION, AND HYBRIDISM.

"For unto every one that hath shall be given, and he shall have abundance; but from him that hath not, shall be taken away even that which he hath." Matth. xxv. 29.

THE amount of Energy in the universe is a fixed quality, which we have reason to believe can neither be added to nor diminished by any power short of a miracle analogous to that of Direct Creation. Hence the failure of all attempts to produce perpetual motion, or in other words, to provide a store of energy which will not sooner or later expend itself. The vitality of animals and plants is only maintained by the nourishment which they derive from air and water, as well as from the earth and its products; and it also appears that every living being has a certain amount of latent vital force, varying in different individuals, and capable of being much modified by surrounding circumstances. But although it may be exhausted faster or slower, its ultimate exhaustion is inevitable, (for its recuperative power is always narrowly limited) and this involves the decay and death of the organism.

The exercise of any organ or faculty is always beneficial within certain limits marked out by the potentialities of the organism itself. Although these limits cannot be exceeded, they are capable of gradual improvement within their possibilities of extension. Thus, the eye might be gradually trained to bear a light which it could not possibly endure at first; or a load may be carried, after repeated trials, greater than could be lifted at the first attempt. But even as the slightest motion involves a certain drain on the vital resources of an organism, so an improvement in any organ implies a demand upon the organism which must be met in one form or another. An animal may excel in either strength or speed, but any abnormal increase in one quality would generally be at the expense of the other. We cannot combine the speed of the race-horse, and the strength of the dray-horse in the same animal. Therefore, although organs or faculties are strengthened by exercise, it is also true that any which are not much used become more or less weakened by disuse, the unused organs receiving *less* than their proper share of the vital energy of the body, and the exercised ones *more*.

It has frequently been remarked that men of great genius often leave no children, or their families soon die out, while in other cases their children show less than average ability. Great physical strength and great intellect rarely co-exist; and the French have a proverb, "As stupid as a dancing-girl." Notwithstanding occasional exceptions, a man who is obliged to work hard at some manual employment, will seldom retain the mental capacity for study. There are even many mental qualities which appear to be somewhat incompatible with each other, and are rarely found united. Thus, it very rarely happens that a poet is also eminent in science. Goethe is perhaps the only remarkable exception; but as remarked in the second chapter, all genius of the highest order is sporadic, and consequently exceptional in its nature.

Some of these illustrations may be susceptible of various interpretations; but they do not affect the main point, *viz.*: that qualities which are exercised tend to increase, and those which are not exercised, to diminish.

May we not legitimately infer that the vital energy has limits to its action; that it works in grooves, and that the deeper channels tend to attract to themselves the energy which would otherwise flow in the shallower ones?* The necessary result is to increase the total amount of differentiation or division of labour, which attains its maximum physically, in the most highly developed animals, and mentally in the most highly civilized communities.

The life of the species corresponds, on a larger scale, to that of the individual, and the same laws of use and disuse will apply to both. Generally speaking, like produces like throughout Nature; but there is at the same time a vast number of differences of more or less magnitude between individuals. The varying constitution of the parents is doubtless one cause of difference in their offspring; but that it is not the only cause is obvious, if only from the fact that twins, though sometimes resembling each other more than other children of the same parents,

---

* On this principle, Darwin's " Law of Correlation of Growth "; Herbert Spencer's " Lines of Least Resistance," and the " Definite Lines of Development," advocated by Asa Gray, Mivart, and others, will admit of a common explanation.

are never *exactly* alike. It may also be remarked that in large families the intermediate children sometimes resemble each other more closely than those which are born in succession. It would be highly interesting to ascertain at what period of fœtal life the special characteristics of the individual are impressed upon it; but it is nearly impossible to answer such a question satisfactorily in the present state of science, which has not yet succeeded in throwing any real light upon the mystery of the origin of life.

The more important differences which we notice in our domesticated animals and cultivated plants, and in local races in a state of nature, are hereditary, and being thus perpetuated, form breeds or races. The instinct of most animals generally leads them to pair with those most like them,* where perfect freedom of choice exists; and the greater their unlikeness, the greater is their unwillingness to pair. Thus races, once formed, tend to become perpetuated and differentiated.

---

* This is not an invariable rule, and the contrary is frequently asserted; but compare Darwin, "Animals and Plants under Domestication," vol. ii. pp. 102—104.

A character which has once appeared in a race, and has, so to speak, been "bred out," by repeated crosses with individuals not possessing it, has always a certain tendency to reappear at some future period in its descendants, even after an indefinite number of generations. This is called "reversion," a term which we shall frequently have occasion to employ.

Thus it is now admitted by all acquainted with the subject, that our domestic pigeons, which vary to such an extent as to exhibit characters unique in the entire Class of birds, have all descended from the wild rock-pigeon. Specimens occasionally appear in all the domestic breeds which reproduce the colouring of the wild species; and this tendency is much increased by crossing individuals of different races.*

The subject of hybridism, alluded to in the last chapter, here demands further consideration. When species are crossed, they are usually sterile; but sometimes produce infertile offspring. Thus the mule, though generally sterile, has occasionally been known to breed, showing

---

* Darwin, "Origin of Species," pp. 177—180. "Animals and Plants under Domestication," vol. i.

that the rule of sterility is not absolute. On the other hand, races of the same species, though widely separated, produce fertile, though often degenerate offspring.

Individuals resulting from the crossing of species, are termed "hybrids," and those produced by the crossing of races, "mongrels." Darwin attributed the degeneracy of half-breeds, to reversion to a more barbarous condition of humanity,* attributable to the crossing; though moral causes must operate as well. Yet close interbreeding is well known to be highly prejudicial, and the crossing of races not too widely separated, is certain greatly to improve the breed of any domestic animal, except so far as it perils its more unstable and most recently acquired characters, both by the direct influence of the cross, and by the impetus which it gives to reversion.

Again, when a species is first domesticated, the change of food and habits not unfrequently affects its constitution so unfavourably as to render it sterile; though when an animal has

---

* "Variation of Animals and Plants under Domestication," vol. i., pp. 46, 47.

become thoroughly domesticated, it will often breed far more freely than in a wild state. Most of our domestic animals have been handed down to us domesticated, from a very early period; and when first semi-domesticated by savages would be almost under their natural conditions, or at least would not suddenly be exposed to such an amount of change of circumstances as to render them sterile in confinement. This semi-domestication having been carried out (as must have been the case) in a manner which would not produce the ordinary primary effect of domestication, *viz.*, sterility, the secondary and converse effect, *viz.*, increased fertility, would in time result. This would not improbably place originally distinct wild species in a position towards each other resembling that of races, rather than that of species; for many of our domestic animals (cattle, dogs, etc.) are perfectly fertile when their most dissimilar varieties are crossed, although they are believed to be descended from several distinct wild species. But if specific differences become plastic under domestication, they can hardly be regarded as absolute barriers between species and species in a wild state. Practically, however, they are so; yet the whole subject of hybridism is as yet so little understood that we cannot affirm positively whether or not the difference between species and races is one of kind, or only one of degree; and although the latter view is still very far from being established, it would be rash to affirm that the former is the only admissible theory.

CHAPTER VII.

HOMOLOGY.

"This mutual relation is like that of the uses of every member, organ, and viscus in the body; and still more closely resembles the co-ordination of the uses of every vessel and fibre in every member, organ, and viscus, where all and each are so consociated that every one regards its own good in another, and thus in all, and all reciprocally in each. From this universal and individual relation they act as one."

Swedenborg, Heaven and Hell, §. 405.

The fundamental unity of plan, running throughout Nature, combined with endless diversity in detail, is one of the chief arguments in favour of Evolution; and the more minutely these affinities are studied, the more striking do they appear; but while on any theory of special creation, they become more and more inexplicable, they nevertheless lead irresistibly to the conclusion that the whole system of Nature is under the control of one infinite Intelligence. Standing at the base of Creation, we find material Nature built up of sixty or seventy so-called "elements," which by combining in different proportions, form compounds of every kind, nearly in the same manner that a whole language is based upon the twenty or thirty letters of its alphabet. Acids and alkalis, stimulants and narcotics, food and poison, and many other wholly antithetical properties, may exist in two substances of absolutely identical composition, except as containing some atoms more or less of a substance in itself perfectly inert. Nay, it is even probable that were our methods of analysis perfect, we might be able to demonstrate that all the elements are ultimately reducible to one primitive form of matter.

Rising in the scale of being, we find that in the lowest living organisms, such as the Rhizopoda, the substance of the body itself performs all the functions of life; and on ascending to the Zoophyta (as the *Hydra*), they devolve almost entirely upon the skin, for when the animal is turned inside out, the lining of the stomach becomes the skin, and the skin becomes the lining of the stomach. In the higher animals and plants, different organs perform different offices; but these still retain a limited power of performing functions which properly belong to others. Thus, if the stems and leaves of plants are kept from the light, they do not develop chlorophyl, whereas roots and tubers exposed to the air, come to resemble stems and leaves, frequently developing chlorophyl (as may often be seen in an uncovered potato) and assuming the functions of those portions of the plant. Even in the highest animals, the functions of different organs are interchangeable to a limited extent.

Again, unity of structure pervades whole Classes, and yet is not absolutely invariable, as might be expected to be the case, if, as is frequently asserted, unity of design were the chief object and meaning of these resemblances. Throughout the great majority of the Sub-kingdom Annulosa, we find a fixed number of somites, whether the body is long or short. The dragon-fly and the crane-fly have no more and the bee and house-fly no less. Huxley ("Anatomy of Invertebrated Animals," p. 398) fixes the normal number of somites in the Insecta and in the higher Crustacea and Arachnida at twenty.

Again, throughout the Amphibia, Vertebrata, and (with some fossil exceptions) also the Reptilia, we find four limbs, normally possessing five digits; and whether the limbs are used for running, leaping, digging, swimming, flying, or grasping, every bone of one animal, no matter how modified, or for what purpose employed, may generally be compared with a corresponding bone in any other. Yet these digits, as in some hoofed animals, are sometimes found in so rudimentary a form that it is difficult to suppose that they can be of any use to the animal.*

Again, in nearly all Vertebrata, whether the neck is as long as in the giraffe, or as short as in the whale, it is composed of seven vertebræ. Yet, this rule admits of exceptions, thus showing that it is not invariable, and that the real explanation of this uniformity is not to be sought for in the direction of the theory of an archetype. The number of the sacral vertebræ is more variable; in man there are five, but in the ostrich, Apteryx, and their allies, from seventeen to twenty.

A stronger instance of the persistence of homological resemblances which appear to answer no good end, is the presence in some animals of rudimentary organs, which are not only quite useless, but absolutely injurious to

---

* This class of argument must be taken for what it is worth, with the qualification that we are continually discovering the use of arrangements in nature which formerly appeared to be purposeless.

G

their possessors, and which appear at first sight to have been given to them for no other purpose than to complete their homological resemblance to some other animal.

It is strange that the Duke of Argyll should write of man,* "In his frame there is no aborted member. Every part is put to its highest use;" for, apart from Embryology, which we will consider in the next chapter, there are in man more than one of the useless and even dangerous structures to which we have just alluded. For example, man possesses a structure known as "the vermiform appendage of the cæcum," which varies considerably in development, and is sometimes entirely absent. It is perfectly useless in man, but represents an important organ in birds and marsupials. Should any hard substance, such as an apple-pip, happen to lodge in it, it has been known to cause inflammation and death; while the appendage is also liable to cause fatal obstruction in some of the other intestines. Granted that such accidents are very rare, is it conceivable that if man had been independently created, an all-wise Creator would have provided him with so absolutely

---

* "Reign of Law," chapter iv.

useless and even dangerous a structure, for the sole purpose of rendering his internal anatomy more like that of a kangaroo! Can anything be more preposterous than such a suggestion? But if man has undergone a long course of Evolution from previously existing forms, some of which may have possessed, as an important part of their structure, an organ which only exists in man as a useless and dangerous rudiment, its presence becomes at once intelligible, and might even excite our wonder and gratitude to the Creator, who has deigned to work upon a method which permits us in some measure to comprehend the marvellous and harmonious series of gradual changes by which all existing organisms have arisen. This is no solitary case; and Darwin remarks, that "not one of the higher animals can be named which does not bear some part in a rudimentary condition; and man forms no exception to the rule."*

Another class of homologies is presented by the organs of the senses in different animals. The ears and eyes of the Cephalopoda are perfectly homologous with those of the Verte-

---

* "Descent of Man, vol. I. p. 17."

brata;* while the eyes of the Annulosa, though analogous in function, and probably fully as efficient† as those of the higher animals, are not yet proved to be homologous with them. It is true that the Mollusca and Vertebrata are believed to have originated from a creature resembling the larval forms of some of the Molluscoida; but this will not account for the close similarity of the sense-organs of the higher forms, nor for the exact correspondence in function between their eyes, and the very differently formed eyes of the Annulosa. On any theory of casual variation, the mathematical probabilities against the occurrence of such correspondences are practically infinite. We are therefore compelled to admit either that the common ancestor of the Mollusca and Vertebrata must have possessed some innate tendency to acquire analogous sense-organs, or that the necessary modifications must have been guided in one direction in these groups,

---

* See Murphy, "Instinct and Intelligence," vol. i. p. 321, and Mivart, "Genesis of Species," p. 74, etc.

† Many observations have been recorded which appear to prove that insects are fully capable of recognising colour as such as well as sounds; but it is certain that the same classes of sounds and colours produce different effects on different animals.

and in another in the Annulosa, by some intelligent* Power in order to produce the required result. In any case, the facts can only be interpreted as due to intelligent supervision; but the secondary laws by which this has acted are at present but little known, though one of them is in all probability the law of correllation of growth.

The eyes of the Vertebrata differ in power in almost every species, being modified according to the peculiar requirements of each. It is believed that those of nocturnal animals have the power of perceiving certain rays of the spectrum which are invisible to ourselves. Lubbock has recently shown that the colour-perceptions of ants are different to and apparently far more extended than our own. I once read of a naturalist who contrived an apparatus by which he succeeded in looking through the compound eye of a dragon-fly, with the effect of producing a multiplied and greatly reduced image of the object observed. It is very doubtful whether this is the normal

---

* We can hardly suppose the existence of an *unintelligent* Power, or a blind force, in such a connection, unless we regard it merely as itself an agent.

power of the dragon-fly's eye, when used by the insect itself; but it would be highly interesting, if it were possible, to make similar experiments on the eyes of nocturnal insects, or on those of insects whose colour-perceptions are different to our own. The eyes of the Vertebrata could scarcely be made the objects of such experiments; but as the visual powers of diurnal insects appear to correspond approximately with those of diurnal vertebrates, it is reasonable to suppose that the eyes of nocturnal animals have also corresponding powers.

Another class of homological resemblances is that which exists between the sexes. It is well known that the special organs of generation exhibit a very close homological resemblance, and that they resemble each other so much in the fœtus, that its sex cannot be determined until a comparatively late period. In fact, it is now ascertained that both sexes (or to speak more correctly, the potentialities of both) co-exist in the fœtus.* Many circumstances favour the belief that the sex of a germ is not in-

---

* Darwin, "Descent of Man," vol. i. p. 207.

herent, but latent, and dependent on causes at present almost unknown. It may here be remarked that no known vertebrate animal is hermaphrodite, except abnormally and imperfectly.

The workers in a bee-hive are undeveloped females, but it has long been known that by a special process of rearing, the larva of an ordinary worker can be developed into a perfect queen-bee. A far more extraordinary assertion has been made by an American lady Entomologist, that the sex of butterflies is determined by the greater or less amount of food supplied to the caterpillars.* Her observations, however, have not yet been confirmed, and as some caterpillars exhibit sexual differences in that state, it would require very careful and prolonged experiments before such a fact could be considered fairly established.

Still less are the secondary sexual characters exclusive attributes of either sex. Darwin observes † " In many, probably in all cases, the secondary characters of each sex lie dormant or latent in the opposite sex, ready to be evolved

---

* Mrs. Treat, American Naturalist, vol. vii.
† "Variation of Animals and Plants under Domestication," vol.ii.p.52.

under peculiar circumstances." But we may go further than this, and assert without fear of contradiction, that every physical or mental characteristic which exists in either sex, also exists (at least in a latent state) in the other.*

It seems that when one sex is fairly established, its usual accompanying characters appear, those of the opposite sex being held for the time in abeyance. But if the proper development of the characters peculiar to the sex of an animal is prevented, the sexual polarity of the organism is always more or less reversed; the check upon the development of the characters of the opposite sex is removed, and they become in their turn more or less developed. According to Mr. E. Saunders, even bees infested by the curious parasite *Stylops*, assume the outward appearance of the opposite sex. (*Trans. Ent. Soc.*, Lond. 1882, pp. 228, 229.)

Confinement has been known to hinder the development of secondary male characters, especially in birds.† There is a mongrel breed of the domestic fowl, called the Sebright

---

* This is proved by the occasional hereditary transmission of diseases necessarily peculiar to one sex, through the other.

† Darwin, "Animals and Plants under Domestication," vol. ii. p. 158.

Bantam, in which the cocks do not possess either combs, or long tail-feathers. The breed is very infertile, but has been perpetuated for more than seventy years. An extraordinary case has been recorded of an old hen of this breed whose ovaries became diseased, when she assumed all the external characters of a fully developed cock; characters which had lain latent in both sexes of the breed for sixty years!

The most constant secondary sexual organs in mammalia are the mammæ, which are in most cases fully developed only in the adult female. These organs are peculiar to the mammalia, the nearest analogical approach to them in other animals being perhaps in pigeons, whose crops undergo a peculiar alteration in both sexes during the breeding season, for the preparation of the food of the young. In mammalia, the mammæ are present in both sexes, and in early life are not more developed in one sex than in the other, while their structure is identical in both. They are very variable in development, and although this is partly dependent on the condition of the reproductive organs, it is by no means wholly so; and apart from the fact of their being often almost equally developed in both sexes among some tribes of Polynesian islanders,* many instances are on record of mammæ having become fully developed, and functionally active, in the healthy and adult male, both in man, and other animals. It is worthy of remark, that in several of these cases, the organ on one side only has become developed, probably owing to some disease or defect on the corresponding side; and that lactation in the male, when it occurs at all, is generally permanent, because it is unchecked by the constitutional changes which take place in the female.†

A converse case to that of the above-mentioned Polynesians, is found among a dwarf race in Madagascar, in which the mammæ are seldom or never developed, even in the female.‡ Very possibly both their dwarfish stature and the non-development of mammæ may be due to one primary cause, perhaps to the whole tribe having formerly been exposed to very severe privations for many generations. But I do not remember

---

* Brenchley's " Cruise of the *Curaçoa*," p. 218

† The flow of milk is said to be rendered permanent in a milch cow by spaying, though the normal effect of such an operation on an animal at another time would be to produce non-development, or atrophy of the mammæ.

‡ Petermann's " Geogr. Mittheilungen," 1871, pp. 142, 143.

to have read of anything similar among other dwarf tribes.

To account for the presence of rudimentary mammæ, and their occasional development in the male, Darwin suggests that the ancestors of the mammalia may have possessed lacteal glands in both sexes, which afterwards became aborted in the males, being no longer required, perhaps in consequence of the females becoming less fertile.* But this case is perfectly analogous to that of the presence of marsupial bones, and sometimes of a rudimentary pouch in the males of marsupial animals; and if Darwin's explanation was correct, we might reasonably expect to find at least *some* species of mammals with fully developed mammæ, and *some* marsupials with a fully developed pouch in both sexes.† It appears more probable that the structure of one sex (and both sexes, as we have seen, exist potentially in the germ)

---

* Descent of Man, vol. i. p. 209—211.

† In the pipe-fishes the males alone are usually provided with a marsupial sac, the walls of which are supposed to secrete a nutritious fluid. Darwin, *op. cit.* vol. ii. pp. 21, 22. In *Alytes* and some other *Batrachia*, it is the male who takes charge of the eggs which are laid by the female, and carries them about, attached to his own body, till they are hatched.

is always more or less correlated in the other, which, considering the very close interdependence of the sexes in all animals, is no more than we might reasonably expect. This view is, indeed, suggested by Darwin himself, who writes: "It is quite possible that as the one sex gradually acquired the accessory organs proper to it, some of the successive steps or modifications were transmitted to the opposite sex."*

Another secondary sexual character in man is the beard. Darwin accounts for its existence in a rudimentary condition, and its occasional development in women, by supposing that the early progenitors of man were bearded in both sexes.† This supposition, and the somewhat roundabout means by which he imagines it to have become rudimentary in women by sexual selection, is scarcely necessary, for the foregoing remarks will equally apply here; and the presence of horns or long hair is very characteristic of the male (though not peculiar to the male in every species) throughout the whole animal kingdom. The development of hair on

---

* *Op. cit.* p. 208
† *Op. cit.* p. 226 ; vol ii. pp. 275—381.

the face is a very variable character in the Quadrumana (being in every case most abundant in the males) and also in the various races of man. The Caucasian race is usually full-bearded, while the African negro* and Mongolian are almost destitute of this appendage. If, as Darwin thinks, the beard has gradually disappeared in women in consequence of men always selecting those with the smoothest faces, it is difficult to understand why this influence has reacted on the males of some races and not on those of others. It is improbable that Natural Selection should have acted in the one case, and sexual selection in the other; and difference of taste among different races is always liable to fluctuations, and would hardly be a sufficiently permanent factor to account for the difference in result. Yet the presence of long hair round the mouth in both sexes in the fœtus† might seem to lend strong support to Darwin's view; but this may, and probably does point to the characters of a very remote

---

* Some of the negro races of Central Africa are bearded, but this may be due to a Semitic admixture. The black Papuan races are, however, more or less bearded.

† Comp. Darwin, *op. cit.* vol. i. p. 25.

ancestry. The comparative anatomy and embryology of man is a subject of the greatest interest, of which we know comparatively little at present, but fuller investigations will doubtless throw a flood of light on many questions of this nature.

Some of the forms of Homology which we have been discussing, and others which might have been mentioned, have been accounted for by supposing that the Creator laid down a uniform plan, upon which all organisms have been constructed. But to carry out such a system to the extent to which Homology occurs in nature would be wholly inexplicable, if not absurd, on the theory that every organism has been created independently for its place in nature; whereas on the theory of Evolution, Homology becomes intelligible, and does not lead us, as in the former case, to form palpably unworthy ideas of the Almighty. Again, this uniformity of plan, though adhered to in many cases where it appears wholly unnecessary, is not always carried out, where no reason can be discovered why it should not be, which still more conclusively displays the fallacy of the whole theory of Special Creation.

CHAPTER VIII.

EMBRYOLOGY.

"My substance was not hid from thee when I was made in secret, and curiously wrought in the lower parts of the earth. Thine eyes didst see my substance, yet being imperfect, and in thy book all my members were written, which in continuance were fashioned, when as yet there was none of them."—Ps. cxxxix. 15, 16.

CLOSELY connected with Homology, and of equal scientific importance, is Embryology. It is now universally admitted that embryological characters are frequently indispensable aids to classification, and are often more useful in indicating the proper position of an animal in the scale of nature, than even a comparison of the adult forms themselves. The simplest of all existing living beings are the Monera,* structureless atoms of sarcode in which even a differentiation between an outer epidermis and an inner nucleus does not appear to exist. From these creatures we ascend in different

---

* A translation of Haeckel's Monograph of this group, by the present writer, will be found in the *Quarterly Journal of Microscopical Science*, for 1869.

directions, and through continually increasing degrees of complexity, to the highest plants and animals which the earth has yet produced.

It is an established law in Zoology that animals within the same Class resemble each other more in embryo, or in their young stages, than in their adult forms, and that an embryo passes through a series of developments which are at first undistinguishable from those of other animals, and it only becomes gradually differentiated from them as its special characteristics successively appear. The differences between the members of any one group essentially consist in the relative development of the elements which all possess in common.*

Since the time of Goethe, it has been generally known that nearly all portions of a plant are simply modifications of the leaves. This was, however, first put forward by Wolff in 1759. Many of the dermal appendages of animals, including even the horn of the rhinoceros, are nothing more than modified hairs.

It was mentioned in the last chapter that the

---

* Compare Carpenter's "Comparative Physiology," ed. 4, p. 101.

sex of an embryo is undistinguishable until a comparatively late period in fœtal life. Further, there is absolutely no means by which a comparative anatomist can distinguish between the fœtus of a man, beast, bird, or lizard, until the limbs have already passed through their earliest stage of development;* and the first rudiments of the legs and wings of a bird are of similar shapes.†

In its earliest stage, it is impossible to tell whether an ovum is that of an animal or a plant; for even the ovum of a mammal presents at one period an extremely close resemblance to *Volvox Globator*, one of the lowest microscopic plants. Next, when the ovum is a little further developed, so as to be recognised as that of an animal, its sub-kingdom is still doubtful; when it manifests itself as a Vertebrate, its Class is still indeterminate; and then the characters peculiar to its Order, Family, genus, species, and individuality appear in succession, though in some cases the specific characters appear before the generic.‡

---

* Von Baer, quoted by Darwin, "Origin of Species," p. 470.
† H. Spencer, "First Principles," p. 336.
‡ Carpenter, *op. cit.* pp. 95, 96.

It was mentioned in the last chapter that many structures which are very important in some animals, exist as useless rudiments in others. This is still more remarkable in Embryology, for teeth which never cut through the gum may be detected in the embryos of whales, and even in those of some birds,* and these absolutely useless organs are subsequently absorbed.

In the human embryo, the brain resembles that of an embryonic fish. The bloodvessels are also formed on the plan of those of a fish; and organs which resemble the permanent kidneys of fishes perform a similar office in the human embryo, until they are superseded, before birth, by the true kidneys.† It must, however, be remembered that an embryo should only be compared with other embryos, and that it would be a mistake to suppose that a human embryo, for example, resembles an adult fish or reptile. The approximate fact is that the embryos, after continuing to resemble each other for a time, finally branch off along

---

* Darwin, "Origin of Species," pp. 482, 483.
† Murphy, "Habit and Intelligence," vol. i. pp. 257, 258.

paths leading to different grades of development.*

The same laws of embryonic development hold good among lower organisms. Thus the larvæ of insects, which may be regarded in one sense as fœtal forms, resemble the early stage of the lower groups of Annulosa, the Annelida, and Myriopoda. The larvæ of all the Crustacea resemble each other very closely on emerging from the egg, but afterwards become differentiated in an analogous manner to the fœtal forms of the higher animals.† The tadpole, which is the larva of the frog, resembles an embryonic fish, and there is a striking resemblance between the Protophyta and pollen-grains.

The law of progressive development in the succession of organic forms on the earth will account for resemblances between the embryos of existing animals, and the adult forms of extinct species. Thus, the Trilobites resembled the larval forms of existing Crustacea, such as *Limulus*; and the most ancient fishes resembled the embryos of existing fish. In the same

---

* H. Spencer, "Biology," vol. i. pp. 143, 144.
† Carpenter, *op. cit.* p. 99.

manner, the *Teleosaurus* resembled an embryonic crocodile.*

On the theory of Evolution, a flood of light is thrown upon these facts by a remark of Darwin's, which does not seem to have attracted the attention which it deserves. " At whatever period of life a peculiarity first appears, it tends to appear in the offspring at a corresponding age, *though sometimes earlier.*"† Assuming that "community in embryonic structure reveals community of descent,"‡ the much simpler forms of ancient organisms seem to have become crystallised in the earlier stages of their modified descendants. Important modifications, according to Darwin's law, would gradually tend to appear at an earlier and earlier period; and the parent forms of the species would thus be thrown further and further back, till they and the earlier modifications were at length only to be traced in the embryonic stages of existing organisms. Each modification occurring at an earlier period would be an advantage to the species, as so much

---

* Carpenter, *op. cit.* pp. 107 & 109.
† " Origin of Species," p. 14. [The italics are ours.]
‡ *Op. cit.* p 481.

more length of life available for further developments.

The young of the Herbivora can run and crop the herbage immediately after birth, and are therefore born with their senses and faculties, far more perfect than those of the young of the higher orders of Mammalia. The Herbivora, are animals of very ancient origin, and must certainly have preceded the Carnivora and Quadrumana; and it is therefore reasonable to suppose that they are at present more highly developed on their own plane than the more modern orders. Thus there is nothing extravagant in the idea of the late J. W. Jackson, that in the future course of human development, men will be born with all their faculties, both bodily and mental, developed to an extent which would appear incredible to us at the present day.* It has been calculated that according to the length of time required by man to reach maturity, the natural length of his life should be 120 years. If this reasoning is correct, we may hope that the actual length of human life (the average of which has been steadily rising in Europe for the

---

* "Man," pp. 87—100.

last two centuries) will continue to increase until it ultimately reaches this limit.

Although records of the original forms of organisms are preserved in their early stages, yet the later developments must necessarily also react upon the earlier, although this may not, with our present knowledge, be observable in the fœtus. As the comparative anatomist can restore the idea of an extinct animal from a mere fragment of a bone, so is it likely that a germ has *ab initio* sufficient individuality to reproduce, in its appropriate matrix and by a process somewhat analogous to crystalisation, its parent form.

We might say much more on the subject of Embryology; but will now conclude with the observation that the whole series of facts which it presents to our observation, though more or less intelligible on the theory of Evolution, are simply inexplicable on that of Special Creation.

CHAPTER IX.

GEOGRAPHICAL DISTRIBUTION.

"It is not wholly the external conditions of light, heat, moisture, and so forth which determine the general aspect of the animals of a country. . . . . There are other causes more powerful than climatal conditions which affect the dress of species."

BATES' *Naturalist on the Amazons*, vol. i. p. 19.

IT was formerly believed that all existing animals were descended from those contained in Noah's Ark, but the idea that the Noachian Deluge (although proved to be the tradition of a real event by the striking resemblance of cognate legends among almost all nations) was universal, has long been abandoned by even the most orthodox of theological geologists, such as Hitchcock and Miller. Upon this admission, of course, the supposed theological necessity for assuming a common centre of distribution for animal life vanishes. Let us now glance at some of the facts of Geographical Distribution, and see whether they agree best with the theory of Evolution, or with that of Special Creation.

First of all, on a cursory inspection of the

productions of the various parts of the world, we find that they fall under five or six great faunas and floras, which appear at first sight as if they might have originated in as many centres.

A naturalist can generally tell at a glance from which continent, and often from what country or even district, any collection of natural objects has been brought.

On descending to particulars, we find that those species which have the greatest mutual resemblance generally inhabit adjacent countries, whereas those which show less resemblance, are brought from regions which are more widely separated. Mountain ranges, and sometimes even rivers, present an almost impassible barrier to the spread of many species. An island near a continent usually possesses very few species as compared with the main land; and these will be nearly all identical, though some will probably present at least well-marked varieties. But an islan~~d~~ the middle of the ocean, which has b~~een sepa~~rated from the main land for a long ~~time (if~~ indeed, they were ever connected), ~~usuall~~y produces still fewer species, nearly all ~~distinc~~h (except those which may happen to have been introduced by man, directly or indirectly) will be found to be peculiar species, resembling no others found on the surface of the globe, although generally exhibiting some degree of affinity with those of the nearest land. Mammalia, except bats, have no ordinary means of crossing a broad sea; and bats are almost the only mammals which are met with in oceanic islands.

Batrachians and their spawn are instantly killed by salt water, and, unless introduced by man, are likewise absent.* Many birds and insects which occur on islands are either wingless or incapable of flight, a singular fact which cannot be without some important signification. Many of the beetles of Madeira are wingless, even among those which are considered specifically identical with winged European species. It must not be forgotten that the development of wings in beetles is sometimes a variable character, even within the same species; but this consideration does not affect the problem to be solved, further than that the fact of a character being variable,

---

* Darwin, "Origin of Species," pp. 424, 425.

renders it more liable to be acted upon by Natural Selection; for the preponderance of wingless beetles in Madeira is far too remarkable to be due to any accidental cause. Darwin has pointed out that it would be an advantage to birds or insects landing on oceanic islands to have their powers of flight either increased or diminished, for such islands are generally of small extent, and liable to be swept by hurricanes, and therefore their winged inhabitants are particularly liable to be blown out to sea. If their powers of flight were increased, they would stand a better chance either of returning to the shore they had left, or of reaching some other land; but if their flight was too weak to give them a reasonable chance of ever reaching land again, when they were once blown away from shore, it would obviously be an advantage to them to become more and more terrestrial in their habits, or even ultimately to lose the power of flight altogether, as they would not then be exposed to the danger of being blown out to sea, and drowned.*

---

* "Origin of Species," pp. 153, 154.

A very narrow sea, if it be also very deep, will sometimes form a well-marked, natural barrier between two totally distinct faunas and floras. Thus, the channel which separates the islands of Baly and Lombock, though only ten miles wide, separates two of the primary Regions of Geographical Distribution, the Indo-Malayan and Austro-Malayan, which are not only far more sharply defined on their boundaries than any other contiguous regions, but are far less closely related, as regards the character of their productions, than Africa and India.*

There is a close connection between the size of animals, and the localities which they inhabit. They are frequently smaller on islands than on continents; and when bred in confinement, are apt to degenerate in size. This is supposed to be due to want of room for sufficient variety in the conditions of life, or to too close interbreeding.†

The inhabitants of towns are usually smaller and weaker than residents in the country, which is largely owing to the absence of

---

* See Wallace's works.
† Murphy, " Habit and Intelligence," vol. i. p. 189.

opportunities for exercise, especially during youth; and subsequently to inheritance. On the other hand, herbaceous plants growing on islands where they have few or no competitors, often tend to assume an arboreal habit.*

Difference in the size of animals of the same species does not always arise from one cause. The insects of Trinidad are frequently smaller than those of Guiana, which may arise from Trinidad being an island; but before concluding that this is the correct explanation, it would be necessary to compare the average size of specimens from the opposite coast of Venezuela also. Contrary to the popular impression, great heat appears to have a tendency to reduce the size of insects, perhaps because the larvæ feed up more rapidly in a hot climate. The tropical representatives of widely distributed genera are frequently inferior in size and beauty to those of temperate climates, and rarely surpass them, for the magnificent productions of the tropics usually belong to groups entirely unrepresented in colder regions. In most cases where Indian insects are also found in Europe or Japan,

---

* Darwin, "Origin of Species," pp. 423, 424.

they are smaller, and an unusually hot summer in England may similarly reduce the size of insects. The European Urus was much larger than the Indian Gaur or Gayal.

The islands of the Galapagos Archipelago, off South America, produce many species of animals and plants which are unknown on the mainland, though all belong to American types; but it is much more remarkable that each island is inhabited by different, though allied, species. The insects of Corsica and Sardinia differ widely from those of the mainland; and many species which occur commonly from Britain to Japan or California, differ less at the extremities of their range, than do their Corsican representatives from Italian specimens. That this variation is chiefly due to long-continued isolation is shown by the occurrence of various intermediate forms in Sicily, which has not been separated from the mainland so long.

If we believe in the doctrine of Special Creation, it follows that although many species have been created to range over enormous tracts of country, yet the existence of a deep sea between two islands, or between an island

and the main land, has led to the separate creation of distinct species to inhabit the opposite coasts. Thus every island of the little Galapagos Archipelago has had a small Fauna and Flora created expressly for it; and a special Fauna and Flora, differing considerably from that of the rest of Europe, has been created expressly for the islands of Corsica and Sardinia, while England, being nearer the French Coast, has only received half the French species, and Ireland only two-thirds of the English; and these islands have scarcely been favoured with any species peculiar to themselves. Who will believe that Infinite Wisdom would have stocked the world in so capricious a manner, if each species had been independently created?

If for the sake of variety, why should not every small district in Europe have its peculiar productions, like Corsica, and Sardinia? Or why should not every island of the great Indo-Australian Archipelago possess a totally distinct Fauna and Flora, whereas it forms only two distinct regions of distribution, divided by a strait but ten miles in width? These questions are unanswerable on any theory involving

Direct Creation, for the whole system of Nature would then become either an unintelligible enigma, or an arrangement expressly contrived to mislead all who seek to inquire into its origin.*

The problems of Geographical Distribution are certainly not without their difficulties, even on the theory of Evolution, but we may expect that these will be lessened in proportion to the gradual increase of our knowledge. It appears that when dominant species have become developed in a particular district, they spread from it, as from a common centre, as far as they can do so unchecked, and branch out ultimately into different varieties.

If a large area favourable for their extension exists, without geographical barriers, they may remain unchanged throughout their entire range, but if a colony should happen to become isolated from the original stock, and shut up in a limited district with different or fewer competitors, it will be more liable to become modified into a new form.

---

* The last proposition is ludicrously illustrated by an old theory that fossils were objects brought by the devil from the moon, to deceive geologists as to the age of the earth.

Previous to the Glacial Period, the Fauna and Flora of the whole world were of a much more uniform character than at present. The preglacial Fauna and Flora appear to have been less completely exterminated or modified in America than in the Old World, which would account for American forms being constituted on lower and older types. Species may remain unchanged for any length of time, if their surrounding conditions remain the same; and Murray believes that California still exhibits a somewhat pre-glacial character.*

The greater part of North America, north of Mexico, appears, however, to have been extensively colonised from Northern Asia, subsequently to the Glacial Period, as many Classes of its productions are so closely related to the great Palæarctic Region that it could not be separated from it as a distinct region, were it not that other Classes present well-marked characters of their own. There is often so great a resemblance between the productions of Europe, Asia north and west of the Himalayas, and America north of Mexico, and west of the

---

* "Geographical Distribution of Mammals," ch. iv. & v.

Rocky Mountains and the Andes, that if we confined ourselves to the consideration of those groups of animals and plants in which this resemblance is most marked, we should be compelled to regard all the various Faunas and Floras as offshoots of one stock, from Ireland to Chili. Of course different regions of this vast region have different characteristics, and the temperate forms are only found under the equator at a great height in the Andes; but the Andes, and further north, the table-land of Mexico form barriers which they do not pass, as we find no trace of them either in Mexico, the West Indies, or South America, east of the Andes. It may here be mentioned, that as 30 per cent. of tropical marine fish are met with on both sides of the Isthmus of Panama, Lubbock thinks it possible that the isthmus itself may be of modern origin.* This suggestion is, however, rendered very doubtful both by the presence of European forms in Chili, and by the purely South American character of the Central American and Mexican Faunæ, which cannot have been derived from the West Indies,

---

* "Prehistoric Times," p. 393.

as Florida is purely North American, and the West Indies are quite as Northern as Southern in their character.*

Australia is far below America in the character of its natural productions. It has been so long separated from America (if indeed it was ever joined to it at all) that it forms, with the adjacent islands, one of the primary biological divisions of the globe. But it is isolated from all the great Continents, and therefore its productions are of a very low type, consisting in mammals, chiefly of marsupials (an Order of which some few representatives are likewise found in America), while it still retains the only living representatives of the transitional order, Monotremata. Consequently we find that the animals and plants of more highly developed countries (rabbits, sheep, thistles, etc.,) thrive in Australia and New Zealand as well as in their own country, and must ultimately come into collision with, and probably extirpate many of the indigenous plants and animals. America

---

* All that is meant here is, that the West Indies possess many strictly American forms that occur in North America. The Europeo-American forms do not reach them.

is less behind the Old World than is Australia, but even in America multitudes of European animals and plants have established a firm footing, while a very much smaller number of American species have become naturalised in Europe.*

The American water-weed, one of the few American plants which have become naturalised in England, is an exception which proves the rule; for it is a water-plant, and the rivers of Europe cannot compare with those of America in size or importance. It is therefore not surprising that a plant which has been trained to the severe competition of river-life in America, should be able to hold its own in the small rivers and canals of Central Europe.

It occasionally happens that the same genus, or even the same species, is met with not in adjacent districts, but in widely separated regions. Mivart lays great stress upon these cases,† and appears to think that the same species may sometimes be developed in different countries.

---

* Compare an interesting article on Imported Insects, and North American Insects in Riley's "Second Report on the Insects of Missouri," pp. 8-15.
† "Genesis of Species,"ch. 7

On the other hand, Murray considers that ch cases may generally be explained by the fo er existence of land-connection between he countries in which these allied forms oc r; nor must the influence of the Glacial Perio be overlooked.

Some cases, as for example, the exis ce of lemurs on distant islands, may possib be explained, as suggested by Darwin,* by eir being the remnants of a once widely distri ted group, which has suffered much extinction, nd is now almost confined to Madagascar, ast Africa, and the Indo-Malayan region, compet ion on islands being always less severe tha on continents. But so many forms are comm to Madagascar and the Indo-Malayan region, hat we can scarcely avoid concluding that a d ect communication really existed between em at some former period. Little as we how at present of the productions of East A ca, many forms once supposed to be peculi to Madagascar, have lately been met with on parts of the coast, between Natal and Zanz ar, and others doubtless remain to be discover..

---

* "Descent of Man," vol. I. p. 202.

A second species of the splendid genus of moths, *Chrysiridia* (*C. Crœsus*, Gerst.), has lately been described from Zanzibar. The only representative of the genus with which we were previously acquainted (*C. Madagascariensis*, Boisdu.) was long considered to be one of the most striking insect forms peculiar to Madagascar.

Such cases as an occasional resemblance between the productions of the Canaries and the East Indies are more difficult to explain. The Red Admiral butterfly (*Pyrameis Atalanta*) is common throughout Europe, North Africa, Northern and Western Asia, and North America, but is replaced in the East Indies, the Canaries, and Madeira by the closely allied *P. Indica*. Although an American species of the genus, *P. Virginiensis*, also occurs in the Atlantic islands, and although *P. Atalanta* has recently been introduced into Madeira, and *P. Indica* into Portugal, yet it seems almost incredible that the latter species should have been accidentally introduced into the Atlantic islands from the East Indies, and not into any intermediate country.

However, if this was an isolated case, we

might fall back upon this explanation in lieu of a better; but the variety of the Large White Butterfly *(Pieris Brassicæ)*, which is found in Madeira, very closely resembles the Himalayan variety of the same insect.

Two other explanations may be suggested. It is possible that in some rare cases, a species may be exposed to such influences as to cause it to vary in nearly the same manner in widely separated countries; or else that such cases represent an old form of a widely distributed species, which has remained unchanged in some portions of its range, while it has become either extinct or modified in other countries.

CHAPTER X.

VARIATION UNDER DOMESTICATION.

"We can anticipate the time when the earth will produce only cultivated plants and domestic animals; when man's selection shall have supplanted 'natural selection,' and when the ocean will be the only domain in which that power can be exerted, which for countless cycles of ages ruled supreme over all the earth."

WALLACE, "Natural Selection," p. 326.

In Nature, variation usually takes place gradually and imperceptibly, or rather, perhaps, we lack both time and opportunity to trace its course, and call many forms species which we should probably class as mere varieties, if we were better acquainted with their history. But in any case, variation takes place much more rapidly, and to a much greater extent among domesticated animals and cultivated plants than among those in a state of nature. Many of the former vary in almost every conceivable direction in the most marvellous manner, and to such an extent that systematic naturalists, and especially botanists, are compelled to leave them almost entirely out of their consideration.

Sheep, for example, though less variable than some other animals, present varieties with no horns, or with two, four, or six; some varieties have a thick fleece, others a huge tail, and others again very short legs. All these varieties, though they pair together readily, and have undoubtedly sprung from one common stock, can be bred true by taking ordinary precautions, although many of them differ so much that they would probably be placed in distinct genera, if met with in a state of nature. But no animals vary so much as domestic pigeons, which are treated of at great length by Darwin in his work on the "Variation of Animals and Plants under Domestication."* Many of the more marked breeds, such as fan-tails, pouters, and tumblers, exhibit characteristics which do not occur in any of the 12,000 species of wild birds.

The advocates of Special Creation have put forward two different hypotheses to account for these and similar facts. They allege firstly, that domesticated species were originally chosen for

---

* Much of this chapter is taken from the above-mentioned work, which is one of the most learned and elaborate of all the writings of Darwin.

domestication on account of their variability; and secondly, that variability is a quality bestowed upon domesticated species expressly to render them useful to man. Both these theories are unsatisfactory, for although the second may be true to a certain extent, yet it only expresses part of the truth. It is almost certain that savages would pay no attention whatever to anything but the immediate use of the plants and animals they first attempted to domesticate. All the most important of our domesticated animals and cultivated plants have been handed down from a very remote period; for in modern times, man has preferred rather to improve those which he finds ready to his hand, than to attempt to turn wild species to account. At a very early date, various animals were domesticated in Switzerland, and both the wild and tame species differed somewhat from existing species and varieties. At that time, too, the domesticated animals varied much less than at present. The cereals and fruits grown by the ancient inhabitants of Switzerland, were also of different and inferior varieties to ours.

As soon as savages began their first rude attempts at farming, artificial selection (at first unconscious, and afterwards methodical) would come into play immediately; but though this has been practised from time immemorial, ages of experience would have been required before high cultivation, weeding out, and skilful breeding could be carried on with anything like our present success.

Nor have the most variable or useful species always been selected for domestication. Many animals will not breed freely in confinement; and unless an animal would do so to a certain extent, its complete and thorough domestication would be impossible.

Nevertheless, in the early ages of the world, animals when tamed, would generally have to find their own living, and would otherwise be exposed to less artificial conditions of life than would be the case now; and thus their tendency to sterility under changed conditions might be gradually overcome.* With respect to plants, Darwin suggests that we probably owe our knowledge of their properties to our predecessors having been forced by famine to attempt

---

* *Vide anteà* pp. 75—77, where this subject has been already discussed in greater detail.

to feed on every substance which could possibly be eaten.

The ass and the cat have been domesticated for many centuries, but vary very little, whereas the bramble, the fruit of which in its wild state is far more palatable than that of the original stock of many of our choicest garden productions, is very little cultivated in our gardens, though one of the most variable of all wild plants. Nor, when the ordinary action of natural laws is sufficient to account for the variation of domesticated species, as well as of those existing in a state of nature, is it reasonable to suppose that extreme variability is *necessarily* inseparable from, and peculiar to, domestication.

It is more probable that every wild species, if domesticated for a considerable time, would prove itself to be far more variable under domestication than in a state of nature. Those domestic animals which do not vary much are those which cannot easily be submitted to restraint (as the cat); those which are kept in comparatively small numbers (as the guinea-fowl); and those which are of too little commercial value for their breeding to be carefully attended to (as the ass).

The great variation which takes place among domesticated species is of incalculable importance to man. It is due entirely to the influence of changed conditions of life, which allow a few years of domestication to produce the effect of centuries of Natural Selection, though less permanently, as the process is too rapid to allow of the tendency to reversion being bred out. Upon this principle depends everything which makes agriculture, farming, gardening, sporting, or any other occupation connected with animals or plants an art or a sciene, rather than a mere routine.

Ponton has objected * that organisms are modified under domestication not for their own benefit, but for man's, and hence that such variation is opposed to the main principle of Natural Selection, which is that every species shall be modified for its own advantage. But he has overlooked the obvious fact that domesticated species are not maintained by their own strength, but by man's, and that therefore the

---

* "The Beginning," p. 401.

one point essential to their perpetuation is that they shall vary in such a manner as to render themselves more useful or pleasing to man, in order to ensure his protection.

Had the theory of Special Creation been true, we could not reasonably have expected that any appreciable improvement or modification would have taken place in species subjected to domestication. But here as elsewhere we perceive the beauty and harmony of the arrangements which allow the earth and its organisms to improve together, and are able to admire some small part of the exquisite laws by which the Universe is governed, and which absolutely dispense with the necessity for any miraculous interposition, which if it existed in Nature, would for ever render the whole system of Nature entirely incomprehensible to us.

CHAPTER XI.

ORIGIN OF LIFE ON THE EARTH.

"The beginning of life and the end of life are both obscure; the middle of life alone is distinct. . . . I am the eternal source of all beings, Arjuna, for nothing whether at rest or in motion can exist without me."

KRISHNA in the *Bhagavad-Gita.*

THE ultimate origin of life ever has been and perhaps will ever remain a mystery to the mind of man; but the idea of the direct interposition of God in the creation of the world has now become so incredible to scientific men, that they sometimes substitute theories almost as absurd as those put forward by their theological opponents. This is much to be regretted, for nothing injures a cause more than arguments which are seen at once to be fallacious, or put forward with a palpable bias, as they are liable to lead us away from the impartial examination of the subject on its own merits; and this is equally true whether the theory which they are intended to support is true or false. And just as the cause of Theology has been injured by the use of ill-advised arguments by its supporters, similar harm is liable to be caused to science by injudicious arguments brought forward in support of doubtful theories.

The double use which may be made of an argument invented simply to explain away facts which we are unwilling to admit, may be curiously illustrated by the now almost forgotten theory which Gosse, in his "Omphalos," dignifies with the title of the "Law of Prochronism in Creation." This "law" sets forth that the Bible states that the earth and its organisms were created less than six thousand years ago, and yet all organisms arise from one another by natural generation in unbroken succession, so that we cannot imagine a seed except as having fallen from a plant, or a plant except as having grown from a seed; nor can we conceive of an egg which was not laid by a bird (or other oviparous animal), nor of a bird which did not spring from an egg. Granting this, it follows that the earth must have been created nearly as it now exists, containing geological formations which were never deposited, filled with the remains of animals and plants, which never really existed, but which would have existed, if the earth had existed longer than the Bible tells us. A century ago, the very existence of fossils was held to furnish unequivocal evidence of the reality of the Noachian Deluge; and Voltaire wished to dispute its occurrence.* He did not, however, suggest some different explanation to account for the presence of fossils, but anticipated the absurd argument with which Gosse, in our own day, thought to bolster up the literal infallibility of Genesis, by suggesting that fossil shells were only appearances, and probably never had any real existence.

One of the recent scientific theories respecting the origin of life, which attracted an unusual amount of attention at the time it was promulgated, was brought forward by Sir W. Thomson in his address to the British Association in 1871.† It may be briefly stated as follows:

The heavenly bodies are moving through space in all directions without any intelligence to guide them [how does he know that?] and

---

* Cf. Miller's "Testimony of the Rocks," pp. 306-310, and Goethe's "Wahrheit und Dichtung," iii ch. 1.

† "Report," pp. civ. cv. [It is now stated that this suggestion was made simply as a joke!]

therefore must sometimes collide. A mass of matter thus detached from an inhabited world with germs attached to it might not experience more violence than a piece of stone dislodged in blasting; and if it fell on an uninhabited world, the adhering germs would in time stock that world with life.

It is true that masses of matter which might possibly be fragments of disrupted planets are scattered through the solar system, and there is reason to believe that collisions between planets, or suns and planets may occasionally occur. We may also admit that aerolites possibly contain organic remains; and that living germs carried to an uninhabited world fitted for their reception would certainly develope themselves. But granting all this, it appears very doubtful whether any living germs could be transmitted by aerolites from planet to planet.

The initial temperature of aerolites is that of the interstellar spaces;* but on striking the

---

* The absolute zero of temperature, or the total absence of all heat, is stated by Murphy ("Habit and Intelligence," vol i. p. 16) to be 492° Fahr. below freezing; but we have no means of ascertaining whether such a temperature has any real existence in the Universe.

K

atmosphere of a planet, they are generally heated to redness, and entirely dissipated; while if they reach the surface of the ground at all, they are generally split into fragments in their descent, and are always too hot or too cold to be touched outside, and if broken open immediately, are often if not always too cold to be touched inside, no matter what may be their external temperature. Surely these violent alternations of temperature would be sufficient to destroy the vitality, if not the very existence of any conceivable germ! Nor, as we have already mentioned, could such adventitiously introduced germs develop themselves, except upon a world already prepared for their reception.

Moreover the organic life of the earth shows no signs of having been derived from any extraneous source. On the contrary, all our knowledge of geology tends to prove that terrestrial organisms, so far as know them, have run a definite course, and that their origin is to all appearance to be sought for in this world alone. It is probable that the inhabitants of every planet are physically complete in themselves, up to the stage of development to which they have attained, and thus that every planet is biologically independent of the others. Miraculous agency in the development of the earth is in the present state of science incredible; and our increasing knowledge of nature always tends to prove more and more convincingly that the origin of life is a natural phenomenon. It therefore follows that the organic world must be evolved from inorganic matter by the regular action of the unchangeable order of Nature. Sir W. Thomson's theory, even if fully admitted, would only account for one link in an endless chain, and could never bring us any nearer to the ultimate origin of life.

Beale, in his "Life Theories and Religious Thought," strongly insists on the difference between living and non-living matter, and asserts that if life became extinct upon the earth, the physical conditions of the world would remain unaltered, but that life could never reappear without the intervention of some power able to overcome ordinary tendencies, and capable of setting at nought natural [physical?] laws. Much of this reasoning may be admitted, for it is now believed by the best authorities that life precedes, and is essentially independent of organisation.* The origin of life is perhaps a problem analogous to the origin of gravitation; and science may ultimately teach us that life is a force pervading all nature, and not merely so-called living organisms. It is, however probable that the destruction of organic life throughout the world would be a break in the chain of causation which would prevent its ever being evolved again under similar forms. When a species once becomes extinct, it can never reappear, for the conditions that led to its evolution can never return; and we may reasonably suppose that the same argument would apply to organic life as a whole.

The Protista form the lowest grade of living beings, and from these all others appear to have arisen; but the main stems are not numerous. Some of these have perhaps branched almost directly from the Protista, as soon as the latter had become differentiated into plants and animals; but in other cases, some of the main stems seem to have been connected very far down in the organic scale. Thus, the structure of the Molluscoida shows us that the Mollusca

---

* H. Spencer, "Biology," vol. i. part 2, ch. 3.

and Vertebrata are probably derived from some primeval form of which the Molluscoida are the modern representatives.

Haeckel has promulgated the theory that the simplest existing group of Protista, the Monera, are continually being produced by spontaneous generation.* But he offers no proof beyond mere conjecture ; and as he has himself observed various modes of reproduction taking place in different species, it is not certain that we have yet actually reached the lowest substratum of organic life. The very lowest forms of Monera, however, possibly multiply by simple fission alone, which is the simplest conceivable form of reproduction.

It has been suggested that the semi-organised mud at the bottom of the deep sea may be the transitional stage between inorganic and organic matter. As life must have originated *de novo* once, it is quite possible that it may do so continually; but it is equally possible that life may have originated only at one particular stage of the earth's development.† Even sup-

---

* Herbert Spencer has explained how simple marine forms may remain unchanged for long geological periods.—" Biology," vol. i. pp. 428, 429.
† Compare Huxley's " Critiques and Addresses," pp. 238, 239.

posing that the conditions necessary for the orderly evolution of life may remain unaltered, still the lowest existing organisms may effectually prevent the development of new forms of life by destroying them in the very course of formation, or as soon as formed. But this is mere conjecture, and it is at present impossible to decide whether life is continually originating on the earth from inorganic matter, or whether it originated only at one particular stage of the earth's development. The probabilities are about equal on both sides, and the experiments by which naturalists have attempted to establish the present origin of living organisms, must be admitted to be extremely inconclusive.

Wallace has suggested [*] that if a mass of matter has sufficient coherence to resist decomposition, and has the power of attracting surrounding substances to itself, we have already a rudimentary vegetable organism; and it is possible that fragments might be broken from the parent mass, and repeat the process.

Such a body might be said to be alive, and to

---

[*] "Natural Selection," p. 272, B.

exhibit a rudimentary form of reproduction, though it would be infinitely removed from even the faintest dawn of conscious existence. Wallace mentions a dew-drop as an illustration of his meaning; a glacier would perhaps be a better illustration; but Murphy, after carefully considering the subject, will not allow that the latter can be termed a living body.* This discrepancy, however, arises only from a different use of terms, and is simply verbal. Wallace has not stated his views on the origin of organic life, but they are probably similar to those which he advocates on the origin of man, which will be considered in our next chapter.

It is possible that the ultimate origin of life, matter, and the immutable laws of nature are under the direct control of the First Cause, and are thus beyond the knowledge or influence of created beings. This view of life seems to be maintained by most writers on the subject, and theoretically even by Darwin himself, though, in practical illustration of his theories, he pushes his reasoning far enough to satisfy even the most materialistic of his followers. Of course

---

* "Habit and Intelligence," vol. I. p. 84.

the Divine control over the laws of nature is more distinctly enunciated by those of Darwin's disciples and critics who have approached the subject from a more decidedly theistic standpoint. But this control does not necessarily imply Creation, in the received sense of the word; which, if matter be eternal, as it very possibly is, can never take place at all.

CHAPTER XII.

COURSE OF DEVELOPMENT ON THE EARTH.

"Heaven has connexion with other worlds. Its inhabitants are God's messengers through the creation. . . . In the progress of their endless being, they may have the care of other worlds."

CHANNING.

FESTUS.　Hath space no limit?
LUCIFER.　　None to thee. Yet if
　　　Infinite, it would equal God, and that
　　　To think of is most vain.

BAILEY'S Festus.

ORGANIC life on the earth has passed through a great series of changes, some very gradual, and others perhaps more rapid; but at each era some natural group has attained its maximum of development in numbers, size, and importance, and has then dwindled into comparative insignificance, to be first equalled and then surpassed by some higher group which was passing through its preliminary stages at the period when the other was dominant. Yet the fundamental resemblance combined with infinite diversity which we can trace not only throughout all existing, but throughout all fossil nature also, is not only inexplicable in the theory of Direct Creation, but leads us at once to ask why, if everything has been created independently, God should have cast all objects into so very few moulds? Especially should we ask this if we believe the assertion so often made, and so ill-grounded, that God created the Universe mainly for His own glory,—a selfish and paltry motive that we are certainly not warranted in attributing to the Almighty. Had the world been created merely for display, we should have no reason to expect any unnecessary unity of plan in nature, but rather that while every organism was fully adapted for its own surroundings, it would be altogether independent in design of every other, existing or fossil. If the doctrine of Special Creation be true, the Argument from Design, the keystone of Natural Theology, and therefore the groundwork of all religion, becomes wholly indefensible.

We know that many species have become extinct, but so complicated are the interactions of one species on another, that the extinction of even one form would necessitate the immediate creation of another to fill its place, provided that species were immutable, and originated by Special Creation. Instead of this, the extinction of a species only disturbs the equilibrium for a time, and its place is filled up not by the creation of a new species, but by the surrounding organisms adapting themselves to the vacuum which its disappearance has caused.

The Theory of Evolution thus represents Nature as a vast self-adjusting machine, upheld by the Divine Wisdom and needing no miraculous interposition or interference with its laws, which are sufficiently perfect in themselves to provide for every possible contingency. We can thus discern far higher proofs of the wisdom of God in nature than if we regard the Universe as a toy, created for a mere temporary purpose, and requiring continual readjustments to keep it in working order at all, even for the short period for which it was intended to exist.

So far as the sciences can show us the laws of Nature, they are immutable and unchangeable, in all worlds, and at all periods. We have every reason to suppose that they were the same a hundred million years ago as to-day, and affect all existing things and beings equally, according to their nature, from the atom to Sirius and Alcyone, or from the moner to the man or the archangel.

Wallace, in his essay on the Limits of Natural Selection as applied to Man,* argues that we may trace in Nature, and especially in the origin of man, the action of intelligences controlling the action of natural law for definite ends.

Startling as it may seem to many that a scientific man should put forward such an idea, it is certainly neither philosophically improbable nor unlikely, unless we venture to deny the existence of any intelligence in the Universe higher than our own. There is really nothing more opposed to natural law in higher intelligences guiding the operation of the ordinary laws of Nature to evolve man, than in man himself improving his domesticated animals in the same manner; and such a theory rests on a very different basis from those which call in the aid of unlimited and gratuitous miracle to reconcile the doctrine of Special Creation with existing facts; as when Mongo Ponton† undertakes to describe the order

---

* "Contributions to the Theory of Natural Selection," pp. 332-372 A.
† "The Beginning," ch. 2

in which the various properties of matter were successively impressed upon it by the Creator.

No doubt many would be inclined theoretically to agree with Channing in his belief that our duties in the future state may include the charge of worlds, who would yet shrink from their practical application as being a literal truth. And yet, why should this appear incredible? Unless we assume that man is the highest being in the Universe, it is all but certain that countless other races of intelligent beings exist, many of whom may be as far superior to man as man is superior to a Moner, and yet so immeasurably below God as to render the distance between themselves and man seem trifling in comparison. Nor can such a belief tend, in the present stage of the world, to Polytheism. Any beings which may be supposed to exist, must necessarily exist only by the will of God, and must exist through, and subject to, the limitations of the laws of Nature, or to speak more correctly, the will of God, as man himself.* Even if we imagine an

---

* "It does not seem an improbable conclusion that all force may be will-force, and thus, that the whole universe is not merely dependent on, but actually *is*, the WILL of higher Intelligences, or of one Supreme Intelligence."
WALLACE, "Contributions," etc. p. 368.

Archangel ruling in Alcyone, and exercising an almost unlimited control over the whole Galaxy, we must yet regard him as controlled by Law as absolutely as the meanest insect.

Man possesses considerable power over the lower animals, and over his fellow-men; and there is therefore nothing unreasonable in Wallace's view that other beings may have power over the development of man, especially when we consider that it cannot be satisfactorily accounted for by the operation of those ordinary physical agencies, which are frequently spoken of as "the blind forces of Nature."

Again, had the world been the direct creation of God Himself, we should expect, and reasonably so, to find it absolutely and unvaryingly perfect in every particular.

This difficulty is so great that it is tacitly implied in the popular theology of the day, which, recognising Special Creation, reconciles it with existing facts by the clumsy expedient of accepting the old Oriental allegory of the "Fall," as a narrative of literal occurrences. But in the place of absolute perfection, which does not exist on earth at all, we find relative perfection, corresponding to the degree of advancement to which the earth has now attained. Absolute perfection must be infinite, and therefore beyond human comprehension, as well as incapable of any further modification or improvement.

Again it may well be doubted whether absolute perfection is logically possible, for any created object must be at least relatively inferior to the Creator; and the wisdom of God is far more manifested in the construction of a world which is ever developing new beauties and uses, and thus ever advancing towards a higher grade of perfection, than if the world had been an unchangeable machine, so perfect in itself that no further improvement was possible or even conceivable. And with regard to the possibility of God being able to create anything infinite, that is, equal to Himself, we may here remark that although some theologians have not scrupled to assert that God could create any number of infinite beings equal to Himself (basing this very doubtful proposition on the totally irrelevant fact that a human son is not necessarily inferior to his father), yet most modern philosophers now believe that the Divine Omnipotence does not extend to a mathematical contradiction, and that God Himself is not able to make two and two either more or less than four.

Darwin insists very strongly upon the imperfection of the geological record, or the history of the world as revealed by its fossils. At first sight this might appear like an *ex parte* statement, but it is merely the statement of existing facts; for the Quaternary fossils lead down to our own day, and yet even these are very imperfectly preserved, while the great breaks and gaps which exist in our knowledge of all the earlier periods, clearly indicate that if the fossils of the Quaternary period, imperfectly as even these have been preserved, form an almost unbroken series leading to existing forms, the history of terrestrial life, if complete, would show that it had run in an unbroken course from its first origin to the present time.

There is no very manifest increase of perfection among established types, for these become as it were stereotyped, and the higher forms which supplant those of a lower type rise not from the latter, but from the lower and more plastic forms below them. For this reason, all the main groups branch off very low down from the common stem. Still, additional forms of existing types are always arising, and increasing differentiation, and consequent improvement is thus secured.

If we survey the progress of Vertebrate animals, we can trace four great periods. First, there was a time when the earth seems to have been almost covered with water, and when fish of all sorts and sizes predominated. Most of these were ganoid fishes, of a type which has now very few existing representatives. When the earth became more solid, an era of reptiles succeeded. The greater part of these died out, and an era of warm-blooded animals followed. Then came the commencement of the Human Period, and man is now more widely distributed and dominant than any other animal or group, recent or fossil.

His present pre-eminence does not correspond with that of any other species, however dominant or widely distributed, but rather with that of a whole Class, an analogy which seems to have been overlooked in most discussions on his position.

In each era the highest animals which the earth was then fitted to sustain have always been dominant, while the inferior groups which preceded them have become much reduced in size and numbers, though not absolutely extinct. There is, however, one argument which may be urged with some apparent force against the continual perfection of development, which is that the existing representatives of fossil species are generally very inferior to them in size, and sometimes in other qualities. Ponton * goes so far as to assert that instead of the strongest forms having prevailed, the powerful animals which were well able to defend themselves have become extinct, while their weak contemporaries have survived to our own day; and to urge this as a conclusive argument against the theory of Natural Selection. It is true that one might as soon compare a cat with a lion, or a badger with a polar bear, as a sloth with a megatherium; and it would be absurd to contend that the smaller animal was the improved descendant of the larger. Still it must be remembered that a dominant group (with the exception of man, in whose case the

---

* "The Beginning," p. 385. He neglects to inform us whether he considers that the strong were destroyed, or the weak preserved by natural or supernatural causes.

species or varieties are very few, and differ but little in size) generally contains a great number of species, varying in size as well as in other particulars. Some writers, however, regard very large size as itself an abnormal character in a group.

A large species must necessarily exist in smaller numbers than a smaller one,' if only because it requires more food, and fewer individuals can obtain food over a given area. A large species, too, is less able to avail itself of shelter, and is consequently more exposed to climatic influences. Again, it is almost always less prolific, and less strong and fierce in proportion to its size than smaller species. All these circumstances give large species a certain disadvantage in the struggle for existence; and hence, when the earth is becoming gradually fitted for a higher order of beings, and unfitted for those which formerly possessed it, the large species are (as a rule) the first to disappear, while the smaller species are able, in some cases, to prolong their existence for ages longer. It is probable, too, that as man was contemporary with the huge mammals of the Quaternary period, some among them, at least, owe their extermination either directly or indirectly to human agency.

Although the crocodiles and great serpents * still survive the far-distant period when the dominant forms of life were Reptilian, the crocodiles (and, indeed, some of the most formidable serpents, such as the Anaconda) belong as much to the water as to the land; and aquatic organisms are all developed on a much lower plane than terrestrial, as we may perceive when we reflect that nearly all the higher groups, both of animals and plants, are terrestrial, or at least amphibious; while many of the lower groups are exclusively aquatic. The iguanodon, and other gigantic land reptiles gave place to the more highly organised Mammalia two geological ages since. Herbert Spencer † says that the crocodile and pike have no definite period of growth, but continue to increase in size as long as they live. He attributes this to their sluggish habits, and to the comparatively small exertion and expenditure of vital energy re-

---

* There is some reason to suppose that serpents, and possibly some forms of four-footed reptiles much larger and more formidable than any now existing, may have survived till comparatively recent historic ages.

† "Biology," vol. i. p. 125, 126.

quired to obtain their food. It is obvious that his remarks will also apply to the boas, pythons, and great tortoises.

We thus perceive that the Human Period is the natural successor of those which have preceded it. The discoveries of science have long been breaking down the barrier which was formerly supposed to exist between the physical nature of man and that of the lower animals. But continued observations have also been breaking down the barrier between Instinct and Reason, which was once deemed still more insurmountable; and it is now no longer possible for man to arrogate to himself the *exclusive* privilege of immortality.

There is reason to believe that the social Hymenoptera, and especially ants, have attained to a high degree of such civilisation as is consistent with their nature. They recognise ants belonging to the same nest after months of absence, although they instantly attack a stranger; and some species keep cattle (aphides, etc.); others, slaves (ants of a weaker species than their own, who have been carried off on predatory excursions); while others devote the greater portion of their time to agriculture, some forming beds of leaves which are supposed to nourish the fungus on which they feed, and others storing up or even cultivating grain.* It may fairly be doubted whether our own employments, even in the most industrious and most highly civilised communities, would appear as rational to beings proportionately larger, as those of ants do to ourselves; especially as their language appears to be tactile, whereas ours is articulate. It is, however, difficult to compare ourselves to animals of so totally different a nature; and we will now consider the intelligence of the higher Vertebrata.

It is commonly asserted that birds build their nests purely by instinct; but the correctness of this assumption was questioned by Wallace in his Essays on Bird's Nests, in his "Contributions to Natural Selection." His reasoning, however, is weaker on this subject than usual, and he has been answered with some effect by Bree.† We are, however, always too apt to assume that animals are incapable of

---

* Compare Belt's account of the Leaf-cutting ants, in his 'Naturalist in Nicaragua," Moggridge's "Harvesting Ants," McCook's "Agricultural Ant of Texàs," &c.

† "Fallacies in the Doctrines of Mr. Darwin," ch. xx.

improvement, although our domestic animals afford us daily proof to the contrary. Man himself, with all his advantages of speech, length of life, hereditary training, etc., which animals do not possess, does not progress so rapidly, especially in a savage state of society, that we could reasonably expect to observe a rapid visible improvement in the intelligence of even our domestic animals. It is enough if they show themselves capable of improvement when the opportunity is given them. In this, as in many other cases, we reason from preconceived data, prejudices, and misunderstood or inconclusive facts; and we cannot be surprised if our conclusions should also prove to be erroneous.

Even the ox, though commonly supposed to be a very unintelligent animal, is trained by the Hottentots to tend cattle and sheep, and is also employed in war. It is, however, said (though on somewhat doubtful authority) that the American bison is so stupid that if a herd comes up with one of its own number who has strayed away, he is liable to be mistaken for an intruder and gored to death. If this be true, it shows a very low amount of intelligence as compared with that of ants; but there is probably no Vertebrate animal (man only excepted) whose intelligence will bear any comparison with theirs.

Among ourselves, the dog is usually considered to be the most intelligent animal next to man; and few will question the general correctness of this opinion, though some authors rate the intelligence of the ape or the elephant as far above that of the dog. But in China, where the dog is reared only for food, the case is very different.*

As Darwin has pointed out,† the dog has learned, since he has been domesticated, to express his feelings with tolerable completeness by barking in different tones. An intelligent dog may be trained to comprehend almost everything said to him in at least two distinct languages. No mammal has yet been trained to imitate the human voice, if we except some well authenticated, though very rare cases, in which a dog has learned to articulate his master's name ‡; but it is perhaps not quite impossible that dogs (for instance)

---

* Darwin, " Animals and Plants under Domestication," vol. ii. p. 220.
† "Descent of Man," vol. i. p. 54
‡ Jesse's " Anecdotes of Dogs."

might be gradually trained to speak, after a long course of careful hereditary training and artificial selection with this express object.

Nor are animals incapable of learning from each other, as well as from man. When cats are reared with dogs, they not unfrequently learn to beg, a habit well known to be hereditary with dogs. Darwin * also says that " there is reason to believe that puppies nursed by cats sometimes learn to lick their feet, and thus to clean their faces ; it is at least certain . . . . that some dogs behave in this manner."

It is commonly assumed that animals are mortal, but that men are immortal, nevertheless many European writers have strongly advocated the immortality of animals; and it even forms a necessary portion of certain systems of European philosophy (that of Allan Kardec, for example) as well as of many Oriental systems. It has always appeared to the present writer that if animals or even plants have no immortality in one sense or other (*what* is quite another question), the doctrine of human immortality becomes utterly

---

* " Descent of Man," vol. i. p. 44.

indefensible. Throughout Nature, we see nothing lost or wasted, and if this is the case with matter or force (the lesser), must it not also be true of life (the greater)? A steam-engine wears out or blows up; there is the iron, but where is the steam? An animal dies; there is the body, but where is the life? To affirm that it has ceased to exist would be as absurd in the one case as the other.

Nevertheless the human mind is evidently so far superior to that of any other of the Vertebrata, as to have led several scientific men belonging to the most opposite schools of philosophy * to regard man as forming the initial type of a new kingdom, as widely removed from the animal kingdom as the animal from the vegetable, or the vegetable from the mineral. As we have seen, his *geological* status is that of a Class.

If the above conclusions are sound, they necessarily involve the pre-existence of the soul; and it is conceivable that an eternal past lies behind us, as well as an eternal future before us. If this be admitted, we may ask whether the

---

* Carpenter and Jackson may be mentioned.

present is our first existence as rational beings; and whether our terrestrial lives (if we have more than one) succeed each other immediately, without any intervals of other-world experience? In answer to the first question, we may say that if there is a succession of lives, it will probably be from lower to higher, corresponding with progressive development on the physical plane; and consequently we may suppose that we have reached our present stage of civilisation by passing through some at least of the lower stages of humanity. The second question may be decided by a mere reference to the vast amount of testimony which all ages have furnished to the existence of disembodied human spirits.

There are not many serious objections to the doctrine of successive existences. One of the most important is perhaps that if we lose our memory, we are practically new creations. But memory is often lost, wholly or partially, in the present life; and in the rare phenomenon of double-consciousness, persons have been known to fall into a deep trance, which has erased the memory of their former life, without at all impairing their reason. A second deep sleep, after an interval, has blotted out the intermediate period, but restored the memory of all that took place before; and these alternations have been known to continue for some years. Even the wild possibility that the same body might be alternately inhabited by two different spirits is negatived by the occasional final blending of these two states.

A second objection is that, assuming a previous life, it would be neither possible nor just that we should be rewarded or punished for the acts of a forgotten existence. But even granting that it is really forgotten (for we have no proof that it is really more than dormant), the old vindictive idea of punishment has long been abandoned by all but theologians; and few will now maintain that punishment ought (at least in theory) to aim at anything but the reclamation of the offender. Consequently, if evil tendencies exist, we must necessarily undergo the treatment required for their cure, whether they are hereditary* or derived from other stages of existence. The latter supposition is equivalent to our being punished in this life for

---

* Vide *posteà*, pp. 183, 184.

the faults of one which we have forgotten; but no needless suffering is permitted in the Universe, and in such cases, we should suffer the needful discipline without its being aggravated by the additional suffering of knowing that we had wilfully brought it upon ourselves.

Finally, I will add that those who look through the teachings of Jesus in the Gospels, will find many passages which appear to imply a succession of existences, in some of which (Matt. xix. 29, etc.), there is a distinction drawn between a future life "in this present time," as opposed to that in "the world to come." The latter is plainly Nirvana, which is not surprising* when we consider the great similarity between Christianity and Buddhism; and the former implies a state of existence similar to the present. It is quite possible that the text, " All who take up the sword must perish by the sword," which is not always literally true in one life, was intended to imply a succession of existences.

---

* In an article in "Human Nature" for June, 1876, I have gone further into the various objections to the doctrine of successive existences, and the passages bearing upon it in the Gospels, than I have thought necessary here.

CHAPTER XIII.

THE DESTRUCTIVE AGENCIES OF NATURE.

Οὐρῆας μὲν πρῶτον ἐπῴχετο καὶ κύνας ἀργούς.
Αὐτὰρ ἔπειτ' αὐτοῖσι βέλος ἐχεπευκὲς ἐφιεὶς
Βάλλ'·αἰεὶ δὲ πυραὶ νεκύων καίοντο θαμειαί.
Il. i. 50-52.

THE end and object of Nature being to people the earth with the highest organisms which it is capable of supporting at each successive period, a continual and very large destruction of species, as well as of individuals becomes necessary; for species pass through their appointed cycle of birth, growth, decay, and death, in a similar manner to individuals. Darwin justly observes that "we may safely infer that not one living species will transmit its unaltered likeness to a distant futurity. And of the species now living, very few will transmit progeny of any kind to a far distant futurity."*

Every species exists merely on sufferance, or

---

* "Origin of Species," p. 524.

in other words, only so long as it can maintain its ground against enemies and competitors. The constant interaction of one species upon another, combined with the gradual but ever varying influence of climatic changes, and altered conditions of life, constantly maintain every animal and plant which does not succumb to them, at the exact point of training required by the exigencies of its existence. The law of physical nature is—let those live that can, and produce a stronger race: and let those die that must, and yield their place to others.

Human agency is at present the most obvious cause of the extinction of species, and necessarily so, although its effect has been exaggerated, for it is not dominant, but local or waning species which are exterminated by man. It was perhaps not human agency which ultimately restricted the range of *Rhytina Stelleri* to two small islands in Behring's Straits, where it was extirpated by man within a few years after the discovery of its haunts. In the early geological ages, the destruction of species of course arose from so-called "natural causes" alone, but man is now so dominant that all other creatures hold a completely subordinate rank in comparison, and must therefore give way whenever his direct or indirect influence interferes to a sufficient extent with the conditions of their life. Thus, an animal may be extirpated either because it is dangerous, or because it is good for food; and if a forest is felled, or a marsh drained, all the plants and animals which require such conditions for their existence must necessarily disappear.

Sometimes species die out slowly and gradually; but when rapid progress is needed, forces appear to be put in action, which work with much greater intensity. At a former period of the history of geology, it was supposed that the earth was periodically devastated by cataclysms; and although this theory has long been abandoned in its entirety, yet there is good reason to believe that volcanic disturbances, floods, changes of climate, etc. were far more frequent and violent at an earlier period than at present; for the further we go back in time, the thinner and more liable to disturbance would be the crust of the earth.

For a long period the climate appears to have been nearly uniform over the whole earth, and the Fauna and Flora presented but slight differences. Possibly the earth was then vivified rather by its own internal heat, than by the heat of the sun. But this condition of things was only temporary. More or less gradually the earth cooled down to the intense cold of the Glacial Period. Then the surface of the globe became divided into climates, and the cold drove everything before it towards the tropics. Whole continents appear to have been depopulated, great numbers of species were destroyed, and nearly all the remainder must have been profoundly modified.* The original forms of many of our existing animals and plants are supposed to be those which now inhabit the Arctic Regions, or the mountains of the Northern Hemisphere, to which they retreated on the diminution of the cold. In North America, it appears that the remains of pre-glacial life escaped into California, and that continent was subsequently re-stocked from Asia-Europe. It is generally supposed that there has been a succession of glacial epochs; and there is at least good reason to believe

---

* Cf. Murray, "Geographical Distribution of Mammals," ch. iv. v.

that the climate of Europe has undergone comparatively rapid changes, and that it was far more severe two or three thousand years ago than at present. Some have even attributed the degeneracy of the inhabitants of Southern Europe to the increasing mildness of the climate.

There is indeed some evidence which might indicate that the climate was less severe in the Middle Ages than at the present time, but it rests chiefly on the former cultivation of the vine in Britain. But there is no evidence to show that the vine flourished more luxuriantly in Britain then than now; and it is very possible that the wine produced was of so inferior a quality that the vine was not worth cultivating in later times, when wine could more readily be obtained from abroad. Even now, a small quantity of grape-wine is produced in the South of England, so that the preponderence of evidence is certainly in favour of the theory that the earth is still gradually recovering from the effects of the last Glacial Period.*

---

* Comparatively slight geological changes would suffice to influence the climate of Europe to a very considerable extent.

The consequences of the great destruction of species at the glacial period, and the profound modification of the surviving species would be permanent. Many of the checks, direct and indirect, which formerly regulated the numbers and modification of the survivors would be removed, and these would assume a more prominent place in Nature than was possible before, and would increase largely in numbers, and would then come into competition with each other, and thus be subjected, already greatly modified and improved, to the powerful action of Natural Selection, which would intensify and permanently fix the changes which they had already undergone. No more active agency for the rapid development of higher organic life than the Glacial Period could well be imagined.

A part analogous to that of the Glacial Period among animals and plants is played by eras of pestilence among man. Pestilence was one of the agencies which was employed to remove the rotting carcase of the Roman Empire, to allow of the development of the nations of Modern Europe. The most terrible of all pestilences in modern time was the Black Death, which swept through the world like a fire in 1348, and broke out again and again during the remainder of the century, depopulating every country in the known world, those least severely visited losing half their inhabitants, and those most so, nine-tenths. There is even some reason to think that it visited North America also. But no sooner had the pestilence subsided, than it was followed by an unusual number of births, and the gaps in the population were soon filled up. Looking at the history of Europe before and since, it appears by no means unlikely that the unparalleled progress of Modern Europe for the last 500 years may be due in great part to the destructiveness of this plague.* Only the strongest and healthiest would have had the remotest chance of surviving so terrible an era, and they would thus become the progenitors of a much stronger race, both in body and mind, especially when their descendants became more numerous, and were again obliged to compete with each other in the struggle for existence. The weaker members of the population being

---

* It may also have assisted to counteract the baneful influence of religious persecutions and celibacy.

eliminated, would leave more scope for the energies of the stronger.

But eras of rest are no less necessary for progress than eras of destruction. If the destructive agencies were always suspended over man, it is needless to say that no advance in civilisation would be possible. Iceland was originally peopled by the boldest and most enterprising of the Norsemen, but the severity of an almost Arctic climate (which appears, contrary to that of Europe generally, to have increased rather than diminished during the historical period), and the tremendous eruptions of Hecla and Skapta Jökul, have broken the spirit of the people, and left them mere strugglers for a bare subsistence. On the other hand, the present rapid progress of Central Europe may be attributed in part to its recent immunity from the plague, which has gradually withdrawn itself more and more for the last two hundred years, and now, even in the East, is far less common or destructive than formerly. This cannot be due solely to more healthy habits of life, but probably to changed conditions of climate or atmosphere, which are less favourable to this scourge.

It thus appears that cycles of comparative quiet intervene between epochs of great destruction (as is notoriously the case with volcanic eruptions and earthquakes) allowing considerable progress to take place in the interval, which, when it slackens, receives a fresh impetus from the new scourges which goad it on at intervals.

Thus, throughout all Nature, we find that death and destruction have no real existence, for life and progress are ever springing from death, stronger and more vigorous than before. In the hand of God we see that the most wholesale destruction of both individual and specific life becomes one of the principal means of covering the earth with new life and beauty. There is no destruction in Nature, for everything is converted to other uses; and no death, for new life always succeeds the old more abundantly. What support then does Nature lend to that materialism which teaches that the body is a mere self-acting machine, and that the consciousness vanishes for ever at death?

CHAPTER XIV.

PROGRESS OF MAN.

"Horskir hrafnar skolu þér
A' hám gálga
Slíta sjónir or,
Ef þú þat lýgr."
FJÖLSVINNSMÁL.

"Not in knowledge only, but in development of powers, the child of twelve now stands at the level where once stood the child of fourteen, where ages ago stood the full-grown man."
TEMPLE, *On the Education of the World.*

THE history of man as a race exhibits some analogy to the system of organic nature. Man being geologically analogous to a Class, the sub-species resemble Orders in their distribution, and the Races, Genera. We find some nations almost destitute of the element of progress, and remaining in an unchanged barbarous or semi-civilised condition for many ages, thus resembling those organisms in which modification proceeds very slowly. Some nations rise to a great height of civilisation, and afterwards, when their mission to the world is accomplished, fall for ever, to be succeeded by others, which although far below the level of their predecessors at the time, are destined ultimately to lead the world up to a higher point of civilisation than it has previously attained. The fall of a highly-civilised but effete nation is really the scattering of the seeds of its civilisation (more or less freed from the vices which formerly accompanied it) broadcast over the world for the benefit of the future. But for the fall of Greece, Rome, and Judæa, and the consequent dispersion of their traditions and literature, we should not have inherited the arts of the first, the laws of the second, and the monotheism of the third; and without this foundation the development of our modern Teutonic civilisation would have been almost impossible.

Bodily modifications of the structure of man (as of other animals) are not very rapid, but the earliest races of men of which any remains have been preserved, were not only lower than the lowest existing savages, but the shape of the bones and the attachment of the muscles more resembled that which is met with in the Quadrumana than is usually the case at the present day. Nor has modification even yet ceased in man. Concurrently with the greater development of the brain, the jaws of the civilised nations are growing shorter, and, as a natural consequence, the wisdom teeth are becoming rudimentary, for while they are always perfectly developed in the negro races, in the white races they have generally only two fangs, are developed late, and lost early.* The resemblance between the children of civilised races and the adults of savage races has frequently been pointed out. The child is affectionate, impulsive, violent, weak in reasoning power, and quick in imitation; while its face is rounder and its nose flatter than in adults. All these characteristics are usually met with among barbarous nations. But children have a great advantage over savages, for they are born into our civilisation, and it is theirs by inheritance. With savages civilisation is an external agency which they can neither conform to nor resist, and which destroys them by the force of its own superiority. It appears that a civilised nation cannot co-exist with a barbarous one, and that the latter will simply die out in its presence, even without any

---

* Darwin, "Descent of Man," vol. i. pp. 26, 27.

attempts at oppression or extermination. In all parts of the world the lower races are giving way before the higher. In Africa alone, where the climate will allow of no permanent white settlements, can the former hold their ground. Nor, philosophically speaking, ought we to regret the inevitable extinction of the lower races of man. Sentimentally it may be regretted; and we must all reprobate the treatment which barbarians too frequently receive from Europeans, but we cannot overlook the fact that no two men, much less two widely different races, are or can be relatively equal.

It is often said that all men are equal in the sight of God; but even this can only be true in the sense that all His creatures are equally His children. It was in this sense that Christ himself observed that although not a sparrow was forgotten before God yet men are of more value than many sparrows; a very suggestive remark of this profound observer of man and Nature.

Even among civilised races, we cannot ignore the wide differences in mind and character between the Romano-Celtic and the Teutonic nations. The former originate, and the latter execute. The seeds of many of the greatest advances in knowledge and intellectual development, though sown among the Latins, have borne no fruit until transplanted to Teutonic soil. I may instance the Reformation, the Circulation of the Blood, and modern Astronomy. The Albigenses, Servetus, and Galileo preceded Luther, Harvey and Newton.

But if there is such a wide difference between the Latins and the Germans, how much greater must be the difference between the mind of a European and that of a Negro! We must not shut our eyes to the plain facts of Nature, and although it is no excuse for treating the lower races with injustice, or for thinking more highly of ourselves than we ought to think, yet we must not remain wilfully blind to the differences between man and man. In our anxiety for justice at the present day we are liable to fall into extremes, and endeavour to remove real injustice and oppression by prematurely attempting to remove restrictions which are really natural and necessary in the present state of Society, in favour of abstract principles. If we had a Republic, and cast lots for our President, on the abstract theory of the equal right of every member of the Empire to govern the whole, what kind of government might we expect! Although this is an extreme illustration, yet it is sufficient to show that in granting equal rights to others (whether subjects or not) we cannot do more than grant *every man an equal right to every office or privilege for which he is fitted.*

It is frequently overlooked in discussions on the rights of man, colonial questions etc., that it is really as wrong to grant privileges to men who are unable to use them, or certain to misuse them, as to withhold privileges from them to which they are fairly entitled.

No fact in Geographical Distribution is more remarkable than the power which the species of the great continents exert over the native productions of islands, as witness the spread of European plants in Australia and New Zealand. On the theory of Special Creation, it would be difficult to explain why the plants of Europe, which thrive so well in these distant countries, should not have been created there also. It is now known that they originated in the countries where they grow, and have had no spontaneous means of passing the barriers which limited their range.

The beliefs of men are regulated by very similar laws. In all cases where community of belief exists, there has been free intercourse, and we never find the same faith growing up among distant and isolated nations. Race is also very closely connected with belief, for religions will only grow upon their appropriate soil.

Nothing is more suggestive than the broad distinction of character existing between the Teutonic and Romano Celtic nations, such near neighbours in Europe, and yet so different. With few exceptions the former are Protestants and the latter Catholics. It is probable that in those parts of Germany which are Catholic, there may be a larger foreign element; and this is certainly the case in the Protestant portions of France and Ireland. We will not speak here of the Sclavonic races, for everything indicates that they are the germs of a civilisation hereafter to be developed,* and their religion rests at present solely on authority, and has not yet become the

---

* Cf. Jackson's "Ethnology and Phrenology."

America, which were to arise from them within a few centuries. Nevertheless Christianity must be held to have triumphed, in this instance at least, not so much by virtue of its truths, as by its corruptions, which brought it down to such a level that the Teutons were able to comprehend it.* Rome has fully understood what we have forgotten; that children and barbarians must be taught by appropriate methods; but she has forgotten that men, and civilised nations, must walk by reason and not by authority, or must at least be able to reconcile the two.

Our missionaries must learn to recognise that the "heathens" have received religions from God suited to their capacity; and that while it is their duty to instil higher views in the place of degrading superstitions, yet they can effect no real abiding good, unless they build on the foundation of truth which these very superstitions contain, and which, once destroyed, can

---

* Just as it was necessary for the pure teachings of Jesus and Mohammad to be corrupted by their successors, before the world could accept them. If their work could have been carried on by John and Ali, instead of by Paul and Omar, both religions would have remained too highly spiritualised to have done the whole work for which they were sent into the world.

never be replaced. Too often they endeavour to dig up everything by the foundations, to rear a flimsy structure on the bare ground in its place. Nor, unfortunately, do missionaries always confine themselves to rooting out what is indubitably evil. In the South Sea Islands they frequently denounce and discourage many of the most harmless or even civilising customs of the natives, such as their love of flowers, thus roughly tearing up the young wheat along with the tares by the handful. Far more serious charges have been brought against some of the missionaries in Africa.

But it may be asked, Why did not God give the "heathens" a true religion? The answer again involves the whole question of Evolution and Special Creation. Their religion is true—for them—for it is the spontaneous outgrowth of their national life. To ask why they have no higher religion is to ask why it is necessary for the blade of corn to grow before the ear is perfected, or why the plants of Europe, which flourish so well in Australia or America when introduced, are not indigenous in the latter countries. And who dares affirm that his own religion is perfect? Even supposing

America, which were to arise from them within a few centuries. Nevertheless Christianity must be held to have triumphed, in this instance at least, not so much by virtue of its truths, as by its corruptions, which brought it down to such a level that the Teutons were able to comprehend it.* Rome has fully understood what we have forgotten; that children and barbarians must be taught by appropriate methods; but she has forgotten that men, and civilised nations, must walk by reason and not by authority, or must at least be able to reconcile the two.

Our missionaries must learn to recognise that the "heathens" have received religions from God suited to their capacity; and that while it is their duty to instil higher views in the place of degrading superstitions, yet they can effect no real abiding good, unless they build on the foundation of truth which these very superstitions contain, and which, once destroyed, can

---

* Just as it was necessary for the pure teachings of Jesus and Mohammad to be corrupted by their successors, before the world could accept them. If their work could have been carried on by John and Ali, instead of by Paul and Omar, both religions would have remained too highly spiritualised to have done the whole work for which they were sent into the world.

never be replaced. Too often they endeavour to dig up everything by the foundations, to rear a flimsy structure on the bare ground in its place. Nor, unfortunately, do missionaries always confine themselves to rooting out what is indubitably evil. In the South Sea Islands they frequently denounce and discourage many of the most harmless or even civilising customs of the natives, such as their love of flowers, thus roughly tearing up the young wheat along with the tares by the handful. Far more serious charges have been brought against some of the missionaries in Africa.

But it may be asked, Why did not God give the "heathens" a true religion? The answer again involves the whole question of Evolution and Special Creation. Their religion is true—for them—for it is the spontaneous outgrowth of their national life. To ask why they have no higher religion is to ask why it is necessary for the blade of corn to grow before the ear is perfected, or why the plants of Europe, which flourish so well in Australia or America when introduced, are not indigenous in the latter countries. And who dares affirm that his own religion is perfect? Even supposing that Christianity is the only true religion, and will ultimately prevail, will not the Christianity of a few centuries hence be as far superior to our own, as ours is superior to that of the Middle Ages? Dare any thinking man assert that he would be able to comprehend the Unveiled Truth in its absolute perfection and purity, even if it were possible for Omnipotence Itself to reveal it to him?

If we would send out missionaries at all, we are bound to select them according to their fitness to teach those whom we wish to instruct. The same man who would do great good among ignorant savages, would be unable to produce any impression at all upon an educated Hindoo or Mohammadan; and instead of propagating his own views, would be liable only to disgust or offend his intended proselyte. Conversely, a man who is competent to argue with educated men of another religion on their own ground, and is able to cause them to respect his opinions (whether he succeeds in convincing them of the superiority of his views, or not), would be thrown away upon the savages whom the other was fitted to instruct.

It was not the simple-minded and enthusiastic

Apostles, but the clever and highly educated Paul who preached Christianity successfully in the most advanced seats of ancient philosophy and civilisation: too successfully, indeed, for his philosophical subtleties, quickly misunderstood and corrupted by ignorance and fanaticism, have been the chief cause of the numerous perversions of Christianity both in ancient and modern times.

Nor must it be forgotten that not only do the religions which we are in the habit of calling "heathen," contain at least some germs of truth, but as no religion can be supposed to be perfect, so those of other nations may at times contain truths new to us, or appeal to God-given faculties unrecognised in our own religion. Many great minds in Modern Europe have pined for the ideal beauty of the ancient Greek religion, which their reason forbade them to accept, but for which the cold religions of the North supplied no substitute whatever.* Surely we may one day hope for a religion which shall combine

---

* The Catholic Religion has also been called the Religion of Poets and Artists; and truly I know of nothing to compare with the exquisite beauty and tenderness of some of the old Catholic legends.

the excellences of all others, and respond to the varying religious instincts and needs of all men—a religion pure and philanthropic as Christianity or Buddhism, as profound as Brahmanism, as fervent as Islam, and as lovely as the Grecian faith; and which, so far from being opposed to Science, shall rest upon foundations in part at least accessible to scientific demonstration. The time may not be yet, but it will surely come; and the highest intelligence of all civilised nations is already slowly and surely converging towards this point.

It has been well said that an argument is not answered till answered at its best; and in highly civilised countries like India, the missionaries rarely came in contact with the real principles of the religions which they would controvert, but merely with exoteric corruptions, which lead them to denounce the religions themselves as hopelessly corrupt. The knowledge of the real doctrines of religions which are professed in countries where only the higher classes are educated, is frequently confined to these, being sometimes unavoidably, and sometimes purposely concealed from all but the initiated. In such cases, it is obvious that converts will be drawn chiefly from the lower classes; occasionally recruited by a sincere, but probably ill-informed convert of a higher status. I have heard a "converted" Hindoo in a Christian pulpit inveighing against his own religion, which it was clear that he had renounced simply because he was a man of too limited an intellect to comprehend it; and, therefore, neither being able to rise to the heights of his own religion, nor sink to its depths, he chose a middle course, and adopted the religion preached to him by the missionaries. I do not blame him, for he was undoubtedly sincere, and chose the best path open to him; but what can we think of the intelligence of a Hindoo who ridicules the mighty mystery of Maya: that everything we see or comprehend is merely phenomenal, and therefore Illusion; a truth which the greatest minds must perceive only the more clearly in proportion to their greatness, and before which the most daring thinkers must bow. Yet such as these are the exceptional converts of whom our missionaries are most proud.

We are now learning more and more that man with all his intellect is subject to the same laws of Evolution as any other creature. We now know that all mental and bodily peculiarities are liable to be inherited, from the most trifling habit to high genius.* The *highest* genius, however, cannot be expected to perpetuate itself directly, but indirectly, by precept and example, for it is an unusual development prophetic of the future capacities of the race; especially when the genius of a whole nation appears to culminate in one man of unapproachable greatness (as in the case of Christ) and then passes away for ever.

Hereditary influence is an immense power, at present scarcely recognised, and which will never perhaps be fully realised ; for it acts in a mysterious, and at times apparently capricious manner, which it is very difficult to explain or understand. Among the lower animals, habits wholly foreign to their wild nature become hereditary under domestication. It is probable that several of the habits of our modern civilisation become hereditary in a similar manner. Girls learn sewing readily, but boys do not. The tendency to speech, as well as accent, has

---

* See Galton's " Hereditary Genius ; " Darwin's "Descent of Man," etc.

become hereditary in children; though language seems to vary too rapidly. But it may be hereditary to a limited extent; and if it were possible to rear two children, born of English parents, of exactly equal capacities, and under exactly similar circumstances, one in France and one in England, it is more than probable that the child who was brought up in England, would learn to speak and read English much more quickly and easily than the other child would acquire a knowledge of French.

At first sight, it appears to be a strange law that many diseases are hereditary; but as good qualities are inherited so are evil, and it is a fearful truth that " the sins of the fathers are visited upon the children," both physically and morally. Yet on looking at the subject more closely, we find that this is really a wise and merciful provision for the good of the human race. Throughout all Nature, the interests of the individual are invariably merged in those of the species, for the ultimate advancement of the latter. By the law of hereditary transmission, all the qualities of the present, whether hereditary or acquired, are liable to be transmitted to the offspring, Now, hereditary diseases are in most cases those which arise from vicious or unhealthy habits of life. These, affecting as they must the whole constitution, are necessarily transmitted, as well as useful or harmless peculiarities, for our modern forms of civilisation do not compel the inevitable destruction of those who are not altogether " the fittest." Besides the moral checks, and motives to improvement of life or habits which such diseases impose, the force of the principle of inheritance is much modified in the human race by precept, example, and the force of circumstances. Here we behold the real meaning of the term " original sin" (whether our present lives have any connection with a forgotten past or not) and it depends solely on ourselves whether we shall free ourselves, as far as lies in our power, from inherited or acquired evils, and transmit a sounder bodily and mental constitution to our children than we have received from our parents, or whether we shall yield to evil, and hand down curses instead of blessings to our descendants.

Many remarkable statements have been published as to the proportions of the sexes being dependent on the requirements of society, even when these are merely temporary and artificial.

It is said that in colonies where the male population is generally in excess, the proportion of female children is very large, until the numbers of males and females is nearly equalised. In Europe the number of male births is always slightly in excess of female, but as the rate of mortality is higher among male children, and as men are to exposed to many casualties from which women are exempt, the actual number of living females is usually in excess of males. It is said that in India the comparative numbers of the sexes differ even in adjacent districts, and without recourse to infanticide, according to the prevalence of polygamy or polyandry among the inhabitants.* It is, however, stated that there is no excess of female births over male in the hareems of Siam.† This may perhaps be explained by the fact that in many Eastern countries it is only the rich who can afford to indulge in polygamy, and their wives are frequently brought from distant countries where polygamy is not general. When such a custom is truly indigenous in a country,

---

* "Illustrated Travels," vol. iii. p. 157.
† "Darwin's Descent of Man," vol. i. p. 303, quoted from Dr. Campbell, "Anthropological Review," April, 1870, p. cviii.

it soon makes its effects felt (as in parts of India) by regulating the proportion of the sexes accordingly.

Some of the opponents of Evolution affirm that we see no signs of advancement in man, but many of degeneracy; and a popular author once called " Genesis and Job, Iliad and Odyssey, Edfou and Karnac, Palmyra and Baalbec," to witness to the mental superiority of his ancestors. But he forgot to establish his own descent from Moses, Homer, Rameses, and Solomon; and if he could not do so satisfactorily his argument fails to carry conviction; while even if he could, it does not follow that those who are less nobly descended are inferior to their ancestors, because he confessed himself to be inferior to his own.

It is true that the ancient civilisations have fallen; many ancient arts and sciences have been lost, and some still remain to be rediscovered; and the crown of empire now rests on other lands than Egypt, Greece, or Rome. But although great genius, which is always sporadic, existed in ancient times, and although great works were then undertaken, which could not be equalled at the present day, yet it is not fair to conclude that there has been no progress. Nor, although Modern Europe may well challenge the comparison, is it fair to compare nations at the highest point of civilisation which they were destined ever to attain, and up to which they had been gradually climbing for thousands of years, with a civilisation which is still growing and flourishing, and is the growth of a very few centuries: we might almost say of a very few lifetimes.

Genesis and the Homeric Poems were based, the one on older documents, and the other on popular songs,; and even if we admit (for the sake of the argument alone) that these works are actually the highest attainments of human genius, yet it is probably because our modern civilisation is unfavourable to the highest literary genius, simply because the immense mass of material before us is at present but half assimilated, and destroys all originality by preventing the concentration of our powers on one point. If we would compare Milton with Homer we ought not to be content with comparing their works, but ask what comparison the life of a wandering bard would bear to the noble life of Milton? But when the discoveries of modern science have permeated our minds as thoroughly as the exploits of the heroic age had permeated those of the Greeks of the time of Homer, we may look for literary marvels which shall dwarf everything which the ancient world could accomplish.

Galton appears on the whole to have somewhat exaggerated the intellectual status of the Athenians, when he says,*

"It follows from all this that the average ability of the Athenian race is, on the very lowest possible estimate, very nearly two grades higher then our own; that is, about as much as our race is above that of the African negro. This estimate, which may seem prodigious to some, is confirmed by the quick intelligence and high culture of the Athenian commonalty, before whom literary works were recited, and works of art exhibited, of a far more severe character than could possibly be appreciated by the average of our race, the calibre of whose intellect is easily gauged by a glance at the contents of a railway book-stall."

Now this reasoning looks plausible enough

---

* "Hereditary Genius," p. 342.

on paper, and yet it can easily be shown to be wholly fallacious. The Athenians flourished before the invention of printing, and although a knowledge of letters was much commoner, and books were perhaps cheaper and more accessible than in the middle ages, yet still neither could have been so universally diffused as at the present day. They had therefore no other means of making themselves acquainted with high-class literature than by the public gatherings to which Galton alludes. In our own times, we prefer to read a book which requires serious attention quietly at home; and to read for amusement rather than for instruction during the hurry and fatigue of travelling. If railway book-stalls had been in fashion in the time of the Athenians we have no reason to assume that the works of Homer and Æschylus would have borne a larger proportion to light literature, than those of Milton and Shakespeare (which are never wholly wanting) do at the present day. Nor can we suppose that all who took part in the Athenian literary assemblies were capable of appreciating the intellectual feast before them. Many, perhaps even the majority, would attend them because it was the fashion; because they wished to obtain credit for being persons of cultivated tastes; because they wished to see some "lion" of the day; or from mere love of novelty, for which the Athenians were always proverbial.* It would be interesting to know how many among ourselves really attend a scientific lecture, or a fashionable church, for the sake of the ostensible object of the meeting, and how many from other reasons. It would certainly not be uncharitable to suppose that the Athenians who attended the fashionable assemblies of the period would do so from motives as multifarious as those which avowedly influence ourselves on similar occasions.

No doubt, if Darwin had advertised a lecture on the Origin of Species at the Crystal Palace, he would have had an enormous audience, but a large proportion would certainly have consisted of persons who were quite incompetent to enter into the merits of the question, and another large proportion would perhaps have been quite indifferent to the subject-matter of the lecture. We must not imagine that such

---

* Acts xvii. 21.

an audience would consist solely of Darwins. The same must have been the case to a very large extent in ancient Athens, or not the grade of intellect but the very principles of human nature would have been totally different. Moreover, Galton has reasoned from the freeborn Athenians alone, who can only be fairly compared with our educated classes, and not with the whole nation, and he will not deny that the ability of our own educated classes is above that of the bulk of the population. It is also doubtful whether he is correct in ranking Socrates and Phidias above all who have succeeded them in Europe. Socrates belonged to a type of Eastern sages who are, perhaps, Semitic rather than Aryan in their genius, and who have no counterparts among ourselves. We cannot fairly compare men of Eastern and Western genius, who have very little in common. But it does not follow that the latter are necessarily inferior. Phidias towers above his contemporaries like Shakespeare among the Elizabethan dramatists; but because modern Europe cannot produce a Phidias (to whom Michael Angelo was probably the nearest approach), it does not follow that Europe is necessarily inferior to Ancient Greece, any more than the England of Victoria is inferior to that of Elizabeth because we cannot now point to a Shakspeare.

But even the ancient Athenians could never pretend to such generally diffused intelligence and morality as we see around us in Modern Europe, much as still remains to be perfected. Greece fell, in spite of her wonderful genius and great philosophers, like Rome, mainly from the absence of settled principles of morality, a defect which is but too conspicuous even in the "Republic" of Plato.

Galton* is eloquent on the injury caused to posterity by celibacy in the Church, but he has missed its real end and object. Paul discouraged marriage because it was a hindrance to missionary work; but celibacy was also a necessary reaction in the ancient Church against the flood of licentiousness which was the destruction of so many of the ancient empires. Celibacy was subsequently extended and encouraged from other motives by the mediæval Church after its original object had been forgotten, and in

---

* "Hereditary Genius."

some measure attained; but even then its perpetuation was of enormous benefit to the human race (as pointed out by Theodore Parker) by preventing the formation of an hereditary hierarchy.

Nor can the stupendous architecture of the ancients be regarded as necessarily a sign of ultra-civilisation. The glory of our modern civilisation is its differentiation and independence, which would render it impossible for any modern Cheops* to concentrate the whole industry of an enslaved empire, for half a century, on the erection of a vast monument which shall stand for all time. The vast cost, and the impossibility of compelling whole populations to slave-labour is one reason why such works are impossible at the present day; and this very impossibility is no sign of our inferiority to the ancients, but rather the reverse. Scarcely a city, much less a mere monument, could be reared by forced labour at the present day, though forced labour was employed, not so very long ago, in the erection

---

*I follow the usual story of Herodotus here, merely in illustration of the argument, without in any way pledging myself to the historical accuracy of the tradition which he records.

O

of both Berlin and St. Petersburg. For this, and for many other blessings, we have mainly to thank the great French Revolution, terrible as it was at the time. Among other benefits, the French Revolution practically abolished torture in Europe. The stake and the wheel were still common punishments in France, up to that time; but during the turmoil of the Revolution, and the years of war which followed, there was no time to torture criminals, and they were simply put out of the way as speedily as possible. This gave time for a new generation to grow up, to whom torture was simply abhorrent, and to Napoleon himself we owe the final abolition of the Inquisition in Spain. We may reckon the French Revolution as the real close of the Middle Ages.

The reign of Solomon was the Golden Age of Hebrew history; and yet Solomon, like a true Eastern despot, commenced his reign with the massacre of his father's personal enemies, and his own rivals, under circumstances of peculiar atrocity. Yet crimes which would now be regarded as likely to draw down the immediate vengeance of Heaven, excited no moral reprobation whatever at that time, and did not interfere in the least with the peculiar favour with which Jehovah was popularly supposed to regard the new monarch.

Justice was utterly unknown in the ancient world. The weak had no protector, and the poor no friend. The vices and cruelties of most of the Roman Emperors are without a parallel in modern days; and Imperial Rome was at times almost hell a upon earth. Christian Constantinople was even worse than Rome, being so utterly base that although Rome could still produce great and noble men, up to the last, the history of Constantinople for many centuries is a dead level of the vilest court intrigues, with scarcely one man of ordinary virtue or eminence rising above the universal baseness. And when the Turks were pouring into this Augean stable, as the Goths had formerly poured into Rome, the last Emperor, Constantine Palæologus, who was a hero, such as Constantinople had not seen for many centuries, could only find a few hundred volunteers who had the manliness to take up arms, and stand by him when the city was in her death-throes. But Constantinople was so debased before that time, that it is only just to the much-maligned Turks to maintain that she can never have been worse since, although the "Christian" Greeks may well have aided to demoralise the Turks, as the "Christian" Romans demoralised the Goths.

Up to the latest period of Roman history, the conquered were almost invariably treated with remorseless cruelty. Scipio wept over fallen Carthage, and showed his sympathy by selling 60,000 men, women and children into slavery on the spot. Xerxes wept at the thought of how soon his vast army must perish, and then lashed the sea because he could not lead them on fast enough to destruction: and these were no unfavourable specimens of great chiefs, among the most civilised nations of the time. Among the barbarous nations of the North, such was their ferocity that it shows itself on any and every occasion, no matter how needless or uncalled for. In one of the ballads of the Edda, quoted at the beginning of this chapter, a porter informs his mistress of the arrival of her lover; and what are the first words that spring to her lips spontaneously?—"The ravens shall tear your eyes out on the gallows if you lie!"

In contrast with this, compare our criminal code with that of James the First, or even George the Third; or compare the past and present criminal codes of the Continent. The total abolition of capital punishment throughout Europe cannot be much longer delayed. Even war is very different now to what it was at the beginning of the century. It is seldom deliberately undertaken by civilised nations except as a dire necessity, and its inevitable evils are mitigated as much as possible, not only by those actually engaged in it, but even by disinterested bystanders. So great at present is the public feeling against war and oppression of every kind, that patriotism itself is often branded with an opprobrious appellation; and any patriot who finds it necessary to engage in even the most just and necessary war, or to use the powers at his disposal with even the appearance of severity, to prevent greater evils, is sure to be misunderstood and denounced. And so anxious are we now to put an end to even the appearance of oppression that there is often danger that great injustice may be done in the attempt to right fancied or exaggerated wrongs, or that the idle and worthless, may be encouraged, and the honest and industrious left to struggle alone against unfair odds. Yet these evils are only temporary and reactionary, and are a sign of real progress. Better indiscriminate charity with all its evils than callousness and indifference.

Individual progress is always in advance of national. In private life, the stage of indiscriminate charity is already left behind ; its evils are fully known and recognised, and avoided, as far as possible. In politics we have just arrived at the stage of indiscriminate charity, in consequence of the national conscience waking to the fact that other nations and communities are entitled to fair play as well as ourselves. Great as are the evils of indiscriminate charity in private life, in politics they are far greater, as witness the present state of Ireland and South Africa. But as this is more clearly seen, the evil will right itself in politics, as has been the case in private life ; and science and history both encourage us to look forward to the future of humanity with confidence, sure that no *really* retrogressive movement is possible.

NOTE to CHAPTER XIV.

RELIGION OF THE SEMITES AND ARYANS: MOHAMMADANISM.

Renau has endeavoured to prove that the Semites have an hereditary tendency to Monotheism, but Max Müller denies it, and proves that most of the ancient Semitic races were Polytheists. But the question can hardly be settled in this summary manner. The Semites, like all other races, were obliged to pass through Polytheism on their way to Monotheism, by a gradual transition; and it must be remembered that the three great Monotheistic religions of the world, Judaism, Christianity, and Islam, are all of purely Semitic origin. Not only have the Aryan races hitherto shown themselves incapable of originating a Monotheistic religion, but Christianity, the only one which ever succeeded in grafting itself on an Aryan stock, soon became encrusted with a vast accumulation of dogmas, mainly derived from Aryan Polytheism, from which the Protestant nations have long been struggling to emerge, and have not yet wholly succeeded.

Arabia at the time of Mohammad was Polytheistic in the same sense as the Roman Church at the time of the Reformation; for a great number of subordinate deities were worshipped, and their adoration overshadowed and obscured that of the Supreme Being. Mohammad was really the Arabian Luther, and Islam no more deserves the name of a new religion than Protestantism, being precisely analogous to the Protestant Reformation of nearly a thousand years later. But the teachings of Mohammad fell on Semitic ground, better prepared to receive a Monotheistic religion than that of Europe; and hence Polytheism was immediately suppressed with surprising rapidity in the East, whereas it not only holds its own, side by side with Monotheism in Europe at present; but even among the orthodox Protestant sects, none have yet attained to the grand and simple Monotheism which is the chief glory of both Judaism and Islam. To such an extent do Mohammadans carry this principle, that they agree with orthodox Protestantism (the most exoteric of all religions) in completely ignoring the Feminine Principle in the Deity.

All Aryan religion is more or less progressive, and can only be imperfectly confined within creeds and articles; but Semitic religion is not so. Mohammad had a great advantage over Luther, for he had full confidence in the divine origin of his inspiration, whereas Luther weakly ascribed much of his own to the devil.* A great Semitic prophet

---

* Compare Howitt's "History of the Supernatural," vol. ii. ch. 5.

stereotypes his teaching on his nation like a seal on molten wax; and it preserves it unaltered until another prophet, either greater than he, or better fitted for the age, rises up to carry forward the work which his predecessor has begun. And whereas the religious teachers of the Aryans reason with the people, and rarely claim for themselves divine inspiration, all the Semitic prophets claim to have received their inspiration direct from God, and teach the people in His name; nor indeed would they be listened to for a moment if they did not.

Some time after the above passage was published in the *Truthseeker*, I was gratified to find that Draper, in his "History of the Conflict between Religion and Science," fully recognises the affinity between Islam and Protestantism; indeed he calls the former "the First, or Southern Reformation." It is indeed time for Western writers to do justice to the great Prophet Mohammad. Islam must be judged by its fruits, and by its best examples, not its worst; by the noble lives of Mohammad himself, of his cousin Ali, and of the sons of the latter; and not by Oriental corruption and profligacy, which have remained but little altered through Pagan, Christian, and Muslim times, and are as much opposed to both the spirit and letter of Islam as to those of Christianity. We do not judge of the religion of Christ by the characters of Henry VIII. or Charles II.; and as little ought we to judge of the religion of Mohammad in an equally unfair manner. The charge of imposture formerly brought against Mohammad is happily almost extinct; but it is in this place sufficient to point out that his nearest and dearest friends (many of them themselves men and women of the very highest character) were always among his sincerest and most devoted followers; and that Mohammad himself always disclaimed the power of working miracles; whereas an imposter, in his position, would certainly not have scrupled to lay claim to such powers; and if devoid of them, would have stooped to jugglery. But Mohammad so strenuously repudiated his possession of supernatural powers, that although endless miracles (possible and impossible) are related of many Mohammadan saints, few or none are related of the Prophet himself.

CHAPTER XV.

HARMONY OF NATURE.

"Organic Nature, then, speaks clearly to many minds of an intelligence resulting, on the whole, and in the main, in order, harmony, and beauty, yet of an intelligence the ways of which are not such as ours. . . . An internal law presides over the actions of every individual, and of every organism as a unit, and of the entire organic world as a whole."

MIVART's Genesis of Species, pp. 238, 239.

OF all external evidences of the wisdom and goodness of God, the harmony of Nature, by which is to be understood its power of universal adaptability to all circumstances, is one of the most striking. Not that harmony is absolutely perfect in every particular, for this would imply an absolutely perfect world, which probably does not exist; but everything is adjusted to its place in Nature in such a manner as best to preserve the stability of the whole fabric. Even variation has its limits, which it cannot transgress. Small animals, as a rule, are far stronger and fiercer, and armed with far more formidable powers than those of large ones, in proportion to their size. Imagine an insect magnified to the size of a large animal. It would be endowed with the speed of lightning, clothed in mail that no weapon could ever pierce, and gifted with sufficient strength to tear up the very mountains. Such a creature would be practically secure from all attack, and its powers of destruction would enable it to devastate the world. Even among the Vertebrata, the ferocity of the larger species is not proportioned to their size, and Wood * remarks that if a mole were as large as a tiger, it would be by far the more formidable animal.

Again, we find that all small animals (which are necessarily exposed to much greater destruction than larger ones, individually, though not specifically,) usually produce a great number of young, while the larger animals bring forth but one or two. It is said that a microscopic animalcule multiplies so fast that it might produce a hundred and seventy thousand billions of young in four days.† The greater liability to injury to which the lower organisms are exposed, is often compensated for by a very short life,

---

* "Natural History," vol. i. p. 424.
† Spencer's "Biology," vol. ii. p. 622.

great powers of repairing injuries sustained, and an almost total insensibility to pain. Even in the higher animals susceptibility to pain only exists as a necessary and indispensable condition to their welfare, it being well known that sensibility is almost confined to the skin and outer portions of the body, the deeper tissues being capable of pain in a much slighter degree.

A familiar illustration of the law of fitness is to be found in the well-known fable of the pumpkin and the acorn. It is true that with certain exceptions* such as the brazil-nut and the durian, fruits large enough to cause serious injury to men or animals by falling from a height, do not usually grow on lofty trees. Even these exceptions hardly invalidate the rule, for the fruits mentioned are actually smaller than many which grow near the ground; and if they bore the same proportion to the plants on which they grow, as a pumpkin or a melon, they would indeed be of gigantic size.

If the Universe were not under omniscient agency, its harmonies would be inexplicable. If we admit the theory of Special Creation, we

---

* Compare Wallace' "Malay Archipelago," ch. v.

shall be compelled to ask why this harmony should not be both absolutely and relatively perfect, and why any discords and defects should exist? for that theory makes the Creator personally responsible for them. But the theory of Evolution accounts for both the harmonies and the discords of Nature, and while reconciling both with the infinite beneficence of the Almighty, involves us in hardly any real moral difficulties.

CHAPTER XVI.

SUMMARY AND CONCLUSION.

"Every step which enables us more truly to interpret the workings of the Divine mind in Nature, necessarily brings us nearer to, and gives us a more intelligent idea of, a Creator."
RILEY, *Third Report on the Insects of Massachusetts*, p. 175.

THE subject of Evolution is so vast that in a short treatise like the present it is only possible to call attention to a few salient points; but it is hoped that enough has been said to indicate the importance of this great law as one of the mainsprings of the Universe, and pre-eminently illustrative of the wisdom and goodness of God in the Order of Nature.

It may here be useful briefly to sum up the line of argument which has been followed, and the conclusions at which we have arrived.

(1.) The speculations of antiquity on matters of physical science are nearly worthless, because most of the ancients possessed too slight a knowledge of the earth and the Universe, to form any adequate conception of their real extent and nature. Hence the theory of Direct Creation was a necessity to their minds, that of Evolution being wholly beyond their means of comprehension.

(2.) The laws of Nature are immutable, and are incompatible with the continual introduction of new species into the world by Direct Creation.

(3.) Evolution is now acknowledged as a universal principle in every department of Science.

(4.) Evolution reveals to us the true System of Nature.

(5.) Natural Selection and the laws connected with it have played a great part in the formation of organic Nature as we now see it.

(6.) Evolution will explain the facts of reversion, rudimentary organs, homology, embryology, geographical distribution, etc., better than any other theory.

(7.) The value of the organic world to man depends entirely on the principle of Evolution, and would be greatly reduced if that of Special Creation were substituted.

(8.) Life appears to have originated on this earth by the action of natural laws, and not to have been introduced from without; and it may have thus originated but once.

(9.) The action of intermediate intelligences in the development of the world is highly probable, and in no way detracts from the absolute supremacy of God.

(10.) Man is but one link in the chain of Nature, and a comprehension of the laws of Evolution is of the greatest importance to him in regulating his future progress.

(11.) Man is immortal by virtue of the inherent indestructibility of the essence of Life itself, and not by any exclusive attribute of humanity.

(12.) Death and destruction have no real existence, being always the divinely appointed precursors of higher life and development.

(13.) The law of Evolution explains both the discords and harmonies of Nature, and reconciles both with the wisdom and goodness of God, whereas the theory of Special Creation would only account for the latter.

(14.) The limits of Variation are such as to preserve a relative equilibrium, and to prevent the destructive agencies from acting in such directions as essentially to derange the existing Order of Nature. Thus:—

(15.) Evolution fully explains the relative perfection which we see around us upon earth; whereas if the theory of Special Creation were true, we should expect to find absolute perfection throughout Nature, and indeed could not account for its absence.

These considerations, among others, have convinced the writer that the Law of Evolution is a paramount necessity in attempting to explain the Order of Nature. But the theory of Special Creation is practically untenable. It will not explain a single phenomenon, and is opposed to all the indications of the System of Nature, so far as we can understand them, as well as to any enlightened view of the nature and attributes of the Deity. Moreover, even were Special Creation true, it would be simply incomprehensible to us, and therefore quite incapable of exciting either our wonder or our admiration as rational beings.

www.ingramcontent.com/pod-product-compliance
Lightning Source LLC
Chambersburg PA
CBHW031813230426
43669CB00009B/1118